Lecture Notes in Electrical Engineering

For other titles in this series go to:
http://www.springer.com/series/7818

Steven Powell · J.P. Shim
Editors

Wireless Technology

Applications, Management, and Security

Springer

Editors
Steven Powell
Computer Information Systems Dept.
California State Polytechnic University, Pomona
3801 W. Temple Ave.
Pomona, CA 91768
USA
srpowell@csupomona.edu

J.P. Shim
Department of Management
& Information Systems
Mississippi State University
Box 9581
Mississippi State, MS 39762
USA
jshim@cobilan.msstate.edu

ISSN 1876-1100
ISBN 978-1-4614-2936-4 e-ISBN 978-0-387-71787-6
DOI 10.1007/978-0-387-71787-6
Springer Dordrecht Heidelberg London New York

Printed on acid-free paper

Springer is part of Springer Science+Business Media (www.springer.com)

Preface

Wireless technology and handheld devices are dramatically changing the degrees of interaction throughout the world, further creating a ubiquitous network society. The emergence of advanced wireless telecommunication technologies and devices in today's society has increased accuracy and access rate, all of which are increasingly essential as the volume of information handled by users expands at an accelerated pace. The requirement for mobility leads to increasing pressure for applications and wireless systems to revolve around the concept of continuous communication with anyone, anywhere, and anytime.

With the wireless technology and devices come flexibility in network design and quicker deployment time. Over the past decades, numerous wireless telecommunication topics have received increasing attention from industry professionals, academics, and government agencies. Among these topics are the wireless Internet; multimedia; 3G/4G wireless networks and systems; mobile and wireless network security; wireless network modeling, algorithms, and simulation; satellite based systems; 802.11x; RFID; and broadband wireless access.

The objective of this book is to explore some of these wireless telecommunications issues in depth. In broad terms the topics covered include the following: wireless communications and network technologies; algorithms, methods, simulation, and software; military applications of wireless communications and satellite applications; RFID, sensor networks, and wireless receivers; and global trends. The book is a collective effort on the part of researchers who participated in the seventh annual Wireless Telecommunications Symposium (WTS) 2007. The Wireless Telecommunications Symposium is an interdisciplinary mobile communications and wireless communications conference which brings together leaders and experts from industry, government, and universities around the world to discuss advances in wireless technology, management, applications, and security. WTS is hosted by the College of Business Administration and College of Engineering at the California State Polytechnic University, Pomona and is co-sponsored by the IEEE Communications Society. The theme of WTS 2007 was *The Future of Wireless Communications*. The WTS 2007 participants, who spoke on such diverse topics as the future of wireless communications research, advances in satellite communications, wireless communications investments and new ventures, mobile wireless services and business, wireless network security, wireless business strategy, and the future of deep space communications, included Internet pioneer Dr. Vinton Cerf , Vice-President and Chief Internet Evangelist at Google; Michael Gallagher, Partner, Perkins Coie LLP and former Assistant Secretary of Commerce for Communications and Information and NTIA Administrator; Richard Lynch, Executive Vice President and Chief Technology Officer at Verizon Wireless; John Muleta, CEO of M2Z Networks and former Chief of the FCC's Wireless Telecommunications Bureau; Kristin Rinne, Senior Vice President – Architecture & Planning at AT&T. Dr. George Rittenhouse, VP – Technology Integration, Bell Laboratories, and Ali Tabassi, Vice President – Technology Development, Sprint-Nextel Corp.

One hundred and thirty-thirty papers were submitted to WTS 2007 from authors representing thirty-one countries. These papers were allocated among seven tracks: Global Trends, Governmental Policies and Cultural Implications; Wireless Services: Business Models, Investments and Ventures, and Market Analyses; Wireless Communications and Network Technologies; Algorithms, Methods, Simulations, and Software; Wireless Communications: Military Aspects and Satellite Applications; RFID and Sensor Networks; and Standards and Platforms. The articles published in this book represent some of the best research presented at the conference.

The book has five chapters covering ***Wireless Communications and Network Technologies***. In the **Optimized Seed Node Locations for Infrastructure Wireless Mesh Networks**, Allen, Whitaker, and Hurley investigate the application of wireless mesh network (WMNs) as a potential solution for addressing the problem of providing cost-effective last mile connections in the case of broadband Internet access. In their study, practical deployment issues of WMN are discussed in the context of developing new protocols for all network layers. Also, specific attention is given to broadband access to business and households using infrastructure WMN technology.

In the chapter on **Maximum-Likelihood Carrier-Frequency Synchronization and Channel Estimation for MIMO-OFDM Systems,** authors Salari, Ahmadian, Ardebilipour, Meghdadi, and Cances suggest an alternative solution of using multiple transmit and receive antennas for spectrally efficient transmission, due to scaling problems associated with increasing network bandwidth. High data-rate wireless access is highly demanded in many practical scenarios. Such transmission methods can be used either to obtain transmit diversity or to form multiple input multiple output (MIMO) channels.

In the **Performance Evaluation of EVRC-Encoded Voice Traffic over CDMA EVDO Rev., A**, authors Li, Vukovic, Filipovich, Fleming, Chan, and Lippman discuss the performance of CDMA EVDO Rev A which is considered an advanced technology version of EVDO Release 0, especially for VOIP application. Through the quantitative evaluations with different real implementation scenarios, the authors examine the characteristics of the technology such as strong long-range dependency of correlation structure of voice traffic, heavy-tailed length distribution of voice traffic, and the tradeoff between initial play-out delay and voice quality. They also discuss the impact of hard handoff and random frame errors on voice quality.

In the chapter on **Modified Max-Log-MAP Turbo Decoding Algorithm by Extrinsic Information Scaling for Wireless Applications**, Taskaldiran, Morling, and Kale discuss the algorithm modification of Max-Log-MAP algorithm, which is to scale the extrinsic information exchange between the constituent decoders. The modification aims to reduce decoder complexity while preserving bit-error-rate (BER) performance of the Max-Log-MAP algorithm in real-time implementation of turbo code. The performance of different methods in choosing the best scaling factor is presented along with the principles of turbo coding.

Authors Holub and Tomiska summarize and compare the findings of subjective experiments testing the influence of delays on the quality perceived by end user in the chapter on **In Delay Effect on Conversational Quality in Telecommunication Networks: Do We Mind?** Explanations are provided for the differences in those results, with new tests being performed and the results of the test compared with existing experiment results.

Five chapters in wireless technology illustrate **algorithms, methods, and simulations.** In **Use of Non-monotonic Utility in Multi-attribute Network Selection**, Bari and Leung discuss the basis for evaluating the appropriateness of multi-attribute decision making (MADM) algorithms, which has been used for network selection decisions in a heterogeneous wireless network environment. The authors suggest new approaches, which is to ascertain the appropriateness of MADM algorithms for network selection and analyze the use of MADM algorithms such as TOPSIS, ELECTRE, and GRA. The findings suggest that GRA provides the best approach in scenarios and propose a need of a novel stepwise approach for GRA that use multiple reference networks.

In **Modeling Cell Placement and Coverage for Heterogeneous 3G Services**, Whitaker, Hurley, and Allen analyze the coverage of CDMA-based systems as used in 3G services. The downlink model, its related terms and parameters are introduced. The authors include sections on test point coverage and cell load calculations as well as an algorithmic approach to modeling service dimensions.

Authors Umlauft and Reichl in their chapter on **Getting Network Simulation Basics Right – A Note on Seed Setting Effects for the ns-2 Random Number Generator** outline numerous limitations to the ns-2 random number generator and give recommendations on how to avoid them. The authors conclude by discussing the impact of currently published works that have used the old method of random number generation on simulated results.

In **Finite Automate for Evaluating Testbed Resource Contention**, Liu proposes a formal testbed abstraction and contention model grounded from automata theory. Such proposition is followed by a deliberation of an empirical testbed featured self-organization, dynamic resource allocation, partition, virtualization, and scheduling.

Authors Tarin, Traver, Marti, and Cardona, in their chapter on **Wireless Communication Systems from the Perspective of Implantable Sensor Networks for Neural Signal Monitoring** experiment a wireless communication testbed that consists of Bluetooth technology and 3G network to generate several findings and pinpoint implementation issues. Recent advances in modern neuro-computing heading toward promising clinical applications of implantable neuronal sensing devices have shown the utmost necessity of wireless communication systems.

Three chapters highlight **military aspects of wireless communications and satellite applications.** In **Performance Analysis of Interference for OFDM Systems**, Luo, Andrian, Zhou, and Stephens analyze the bit error rate (BER) of various types of interference for orthogonal frequency division multiplexing (OFDM). The analysis is presented in two formats: analytical and software simulation. Types of interferences covered include barrage noise interference (BNI), partial band interference (PBI), and multi-tone interference (MI). The authors propose two novel interference injecting methods.

In **AVQRED in Satellite Networks,** Byun and Baras examine queuing behavior in satellite networks and propose a new method for overcoming issues which arise from the application of existing AQM methods. The previous 10 years have seen exponential increases in Internet traffic. As a result, active queue management (AQM) has become a heavily studied topic by computer science researchers.

RFID, Sensor Networks, and Wireless Receivers are examined in three chapters. In **RFID Indoor Tracking System based on Inter-tags Distance Measurements,** Bekkali and Matsumoto develop RFID for an indoor tracking system for mobile computing applications used to track distance and location of products. Using WLAN technology, RFID can achieve spatial location information similar to information provided by global positioning systems, but for indoor product location.

Wang and Tang, in their chapter on **Topology-Based Routing for Xmesh in Wireless Sensor Networks** review topology base routing schemes for Xmesh. Xmesh is a multi-hop routing protocol for distributed routing processes, such as those in wireless networks. A detailed simulation and analysis of the routing protocol indicates that average path lengths decreased when a virtual topology was imposed.

In **Efficient Structures for PLL Loop Filter Design in FPGAs in High Datarate Wireless Receivers – Theory and Case Study**, Linn focuses on deliberating an improved design of digital loop filters for phase lock loops in high-speed wireless receivers. The author attributes such saving to the multipliers that are implemented as state machines which compute and sum the partial products iteratively. Significant savings in resources utilization (71–76%) can be achieved provided that certain conditions are satisfied.

In **News Corporation: Facing the Wireless World of the 21st Century,** Cossiavelou and Bartolacci conduct a case study on News Corporation, which utilizes its joint ventures, strategic business planning, and its willingness to embrace the emerging wireless technologies, attempting to establish the notion that News Corporation will appear as a role model in the media industry landscape.

We would like to extend our appreciation to our colleagues on the WTS Committee, especially Tom Ketseoglou, WTS Assistant Chair, Mike Bartolacci, and Katia Passerini, WTS 2007 Program Committee Co-Chairs, WTS 2007 Track Chairs Hussain Al-Rizzo, Jan Holub, Rose Hu, Izabella Lokshina, Santosh Nagaraj, Ehsan Sheybani, Roger Whitaker, and Qing-An Zeng, and WTS 2007 Administration and Operations Chair Steven Curl for their suggestions and help organizing WTS 2007 and to the many reviewers for evaluating the papers submitted to the conference. We also would like to thank our current and former doctoral students, Aaron French, Chengqi Guo, and Jongtae Yu at Mississippi State University for their assistance. Finally, we would like to extend our gratitude to the people at Springer – Jason Ward and Caitlin Womersley – for their help preparing this book.

Pomona, CA, USA Steven R. Powell
Mississippi, MS, USA J.P. Shim

Contents

Contributors

S. M. Allen School of Computer Science, Cardiff University, Queen's Buildings, 5 The Parade, Cardiff CF24 3AA, UK, stuart.m.allen@cs.cardiff.ac.uk

Mahmoud Ahmadian Faculty of Electrical Engineering, K. N. Toosi University of Technology, Tehran, Iran, mahmoud@eetd.kntu.ac.ir

Jean H. Andrian Department of Electrical and Computer Engineering, Florida International University, Miami, FL 33174, USA, Jean.Andrian@fiu.edu

Mehrdad Ardebilipour Faculty of Electrical Engineering, K. N. Toosi University of Technology, Tehran, Iran, mehrdad@eetd.kntu.ac.ir

John S. Baras Institute of Systems Research, University of Maryland at College Park, MD, 20742-5141, USA, baras@isr.umd.edu

Farooq Bari Department of Electrical & Computer Engineering, The University of British Columbia, Vancouver, BC, Canada V6T 1Z4, farooq.bari@att.com

Michael R. Bartolacci Associate Professor of IST, Penn State University – Berks, Reading, PA,US, mrb24@psu.edu

Abdelmoula Bekkali Graduate School of Global Information and Telecommunication Studies, Waseda University, 1011 Okuboyama Nishitomida, Honjo, Saitama ken, 367-0035, Japan, bekkali@fuji.waseda.jp

Do Jun Byun Institute of Systems Research, University of Maryland at College Park, dbyun@hns.com

Jean-Pierre Cances University of Limoges, Ensil-Geste, Parc ESTER, Limoges, France, e-mail:cances@ensil.unilim.fr

N. Cardona Institute for Telecommunications and Multimedia Applications, Technical University of Valencia, Valencia, Spain

Eric Chan Network Advanced Technology, Motorola Inc., Arlington Heights, IL, USA 60004

Vassiliki Th. Cossiavelou Communication Secretary A', Ph.D. Candidate, Aegean University, Greece, v.cossiavelou@ct.aegean.gr

Igor Filipovich Network Advanced Technology, Motorola Inc., Arlington Heights, IL, USA 60004

Phil Fleming Network Advanced Technology, Motorola Inc., Arlington Heights, IL, USA 60004

Jan Holub Department of Measurement K13138, FEE CTU Prague, Technicka 2, CZ 166 27 Prague 6, Czech Republic

S. Hurley School of Computer Science, Cardiff University, Queen's Buildings, 5 The Parade, Cardiff CF24 3AA, UK

Izzet Kale Applied DSP and VLSI Research Group, Department of Electronic Systems, University of Westminster, London, W1W 6UW, United Kingdom, kalei@wmin.ac.uk

Victor C. M. Leung Department of Electrical & Computer Engineering, The University of British Columbia, Vancouver, BC, Canada V6T 1Z4, farooqb,vleung@ece.ubc.ca

Fulu Li The Media Laboratory, Massachusetts Institute of Technology, Cambridge, MA USA 02139, fulu@mit.edu

Yair Linn Universidad Pontificia Bolivariana, Bucaramanga, Colombia, yairlinn@gmail.com

Andrew Lippman The Media Laboratory, Massachusetts Institute of Technology, Cambridge, MA USA 02139

Lei Liu 3738 Evangelho Cir. Sane Jose CA USA 85149, lei.liu@sun.com

Jun Luo Department of Electrical and Computer Engineering, Florida International University, Miami, FL 33174, USA

P. Martí Institute for Telecommunications and Multimedia Applications, Technical University of Valencia, Valencia, Spain

Mitsuji Matsumoto Graduate School of Global Information and Telecommunication Studies, Waseda University, 1011 Okuboyama Nishitomida, Honjo, Saitama ken, 367-0035, Japan.

Vahid Meghdadi University of Limoges, Ensil-Geste, Parc ESTER, Limoges, France, meghdadi@ensil.unilim.fr

Richard C.S. Morling Applied DSP and VLSI Research Group, Department of Electronic Systems, University of Westminster, London, W1W 6UW, United Kingdom, morling@wmin.ac.uk

Peter Reichl Telecommunications Research Center Vienna (FTW), Donau-City-Str. 1, A-1220 Vienna, Austria

Soheil Salari K. N. Toosi University of Technology, Faculty of Electrical Engineering, P. O. Box: 16315-1355, Tehran, Iran, salari@eetd.kntu.ac.ir

James P. Stephens, Sr. Air Force Research Laboratory, 2241 Avionic Circle, N3-F10 Wright-Patterson AFB, OH 45433-7333, USA

K. Wendy Tang State University of New York at Stony Brook, New York, USA, wtang@mail.ee.sunysb.edu

C. Tarín Institute for Telecommunications and Multimedia Applications, Technical University of Valencia, Valencia, Spain

Mustafa Taskaldiran Applied DSP and VLSI Research Group, Department of Electronic Systems, University of Westminster, London, W1W 6UW, United Kingdom, m.taskaldiran@wmin.ac.uk

Ondrej Tomiska Department of Measurement K13138, FEE CTU Prague, Technicka 2, CZ 166 27 Prague 6, Czech Republic

L. Traver Institute for Telecommunications and Multimedia Applications, Technical University of Valencia, Valencia, Spain

Martina Umlauft Women's Postgraduate College for Internet Technologies, Vienna University of Technology, Favoritenstr, 9-11, A-1040 Vienna, Austria, umlauft@big.tuwien.ac.at

Ivan Vukovic Network Advanced Technology, Motorola Inc., Arlington Heights, IL, USA 60004

Lei Wang State University of New York at Stony Brook, New York, USA, leiwang@mail.ee.sunysb.edu

Roger M. Whitaker School of Computer Science, Cardiff University, Queens Buildings, 5 The Parade, Cardiff, CF24 3AA, U.K., r.m.whitaker@cs.cf.ac.uk

Chi Zhou Department of Electrical and Computer Engineering, Illinois Institute of Technology, Chicago, IL 60616, USA

Optimized Seed Node Locations for Infrastructure Wireless Mesh Networks

S.M. Allen, R.M. Whitaker and S. Hurley

Abstract Infrastructure wireless mesh networks offer an elegant potential solution to the last mile problem in providing widespread broadband access, yet there remain significant challenges in deploying such networks in a cost-effective manner. Each subscriber in the network is able to forward traffic to/from their near neighbors, hence providing additional coverage. Here we consider the relationship between consumer demand and coverage, together with the requirements for dedicated network infrastructure. The simulations of consumer subscription decisions presented highlight the importance of considering the take rate for the service in evaluating the potential network performance. This motivates a multi-objective optimization algorithm to determine the optimal locations to deploy *seed nodes* in order to improve coverage with minimal investment.

1 Introduction

Solving the problem of providing cost-effective last mile connections is crucial if broadband Internet access is to be available to all. Recently [1], wireless mesh networks (WMNs) have emerged as a potential solution. WMNs are multihop wireless networks; when source and destination nodes are outside direct communication range, intermediate nodes are used to forward packets between the two. Hence each node in a WMN acts as both a client (the source or destination of traffic) and a router for traffic to/from neighboring nodes. Since they require little or no central infrastructure (depending on the application), WMN are cheap to deploy. New nodes joining a network require minimal installation, so deployment is very flexible. Since there are typically multiple paths between source and destination, WMNs have a natural resilience to node and link failure.

S.M. Allen (✉)
School of Computer Science, Cardiff University Queen's Buildings, 5 The Parade, Cardiff CF24 3AA, UK
e-mail: stuart.m.allen@cs.cardiff.ac.uk

S. Powell, J.P. Shim (eds.), *Wireless Technology*, Lecture Notes in Electrical Engineering 44, DOI 10.1007/978-0-387-71787-6_1,

Practical deployments of WMN raise many interesting research challenges in developing new protocols for all layers. The application of WMN to nomadic and mobile scenarios (such as public safety networks in emergency areas, transportation information systems, and ad hoc networks) has been widely studied (see [2] for a survey of proposed system architectures, applications, protocols, and open research issues). Here we are interested in the use of infrastructure WMN to provide broadband access to businesses and households. For this purpose, WMNs appear to offer an attractive alternative to xDSL/cable and conventional point-to-multipoint (PMP) fixed wireless access schemes, particularly in rural or suburban areas. The advantage to the provider of such a service is the lack of significant up-front investment and increased ease and flexibility of deployment, where the limited configuration required means networks can be deployed incrementally. Initial systems have been proposed and implemented using a range of technologies and frequency bands, ranging from systems making use of directional antenna in dedicated 28 GHz bands (for example [3]) to solutions using 802.11 or 802.16 equipment in unlicensed bands (for example [4]). Networks using propriety technology in licensed spectrum have greater control of the interference environment and can make use of more sophisticated protocols/algorithms for network management. However, this advantage comes at a large expense in hardware costs, hence here we will focus on 802.11/802.16 type WMNs, where there is an abundance of cheap equipment available.

If such infrastructure WMNs are to be widely successful in providing broadband access, it is vital that they achieve the high levels of coverage that exponents of mesh systems predict. Although there are many emerging system vendors[1], start-up network operators and community mesh networks[2], there is little published research concerning their effective deployment, particularly on a wide scale. From a network operator's (either commercial or community) perspective, the key objectives to be considered are providing high coverage to potential subscribers, offering high capacity to existing subscribers, and minimizing infrastructure costs. These multiple objectives should be optimized simultaneously during deployment yet may be in direct competition with each other (i.e., optimizing one means another must be degraded).

Since each subscriber provides coverage to their local area, the overall coverage increases as more subscribers join the network. The converse of this is that coverage can be poor or patchy when there are low numbers of subscribers, which will be the case during the initial deployment of the network. To overcome this limitation in a practical deployment, *seed nodes* (additional infrastructure added purely to provide additional coverage) may be required to be installed. Although there is very little published research concerning these issues, there are some notable contributions. In addition to describing potential mesh architectures, [5] compares the coverage of WMN favorably to point-to-multipoint architectures but notes the requirement for

---

[1] For example, http://www.tropos.com,http://www.motorola.com/mesh, www.meshdynamics.com
[2] For example, socalfreenet.org, www.roofnet.net

seed nodes during the initial phase. Bounds on the capacity of WMN are derived in [6] which uses the idea of collision domains to identify traffic bottlenecks in the network based on the locations of active nodes. All three objectives listed above are combined in [7] which proposes algorithms to satisfy coverage and capacity constraints while minimizing the number of gateway nodes deployed. Other optimization techniques have been proposed to improve the performance of WMNs during their operation rather than deployment, but have mainly focused on considering routing, scheduling, and channel assignment (for example, [8, 9]).

In this work we consider the provision of seed nodes during the deployment of *greenfield* WMN (where there are no existing subscribers), while addressing two shortcomings that are overlooked in existing work. First, in a practical deployment, the location of subscribers is unknown in advance. By considering the probability that a node will wish to subscribe, we examine the expected coverage of the network. The probability of subscription will vary for individual users based on many factors such as the quality of service offered, subscription prices, availability of competing services, and demographics. Second, existing approaches have ensured that coverage and capacity constraints are satisfied for all subscribers while minimizing the number of seed nodes. In practice it is not ideal to prioritize the objectives in this way. For example, it is unlikely to be cost-effective to cover *all* potential subscribers in a region, as ensuring coverage for some individuals at the extremities of the network may require prohibitive levels of infrastructure. This initial work addresses these two issues, by formulating the seed node location problem as an optimization problem with two objectives, with the aim of maximizing the expected coverage (under uncertain subscription), while minimizing the cost of seeds deployed. Section 2 formally describes the network model and the resulting optimization problem, with a proposed algorithm to produce solutions is described in Section 3. Experimental results are provided in Section 4 that demonstrate the relationship between coverage, investment in infrastructure, and the level of consumer demand for the service. The final section draws some conclusions and proposes further work that should be investigated.

2 The Seed Node Placement Problem

The overall aim of the seed node placement problem presented here is to determine locations at which to deploy seed nodes in order to optimize the competing objectives of minimizing infrastructure cost, while maximizing the expected coverage of the network.

2.1 Network Model

We define a WMN to be composed of a set of wireless nodes, each belonging to one of three classes, namely *user*, *Point-of-presence* (POP), and *seed*. For all three classes each node has an associated location.

Users: Let $U = \{u_1, u_2, \ldots, u_N\}$ be a set of N nodes representing the *users* (or potential subscribers, households) in the network region. For a user $u \in U$ let $p_s(u)$ denote the probability that they wish to subscribe to the service. In practice, individual users or subsets of users will have different subscription probabilities based on the demographics of the region or the availability of competing services.

POPs: Each *POP* represents a gateway node, being a connection of the network to the Internet. We are interested in infrastructure WMN to provide broadband access, hence all traffic will be between individual user nodes and a POP; there is no peer-to-peer traffic within the network. Let $\Gamma = \{g_1, g_2, \ldots, g_P\}$ be the set of POPs in the network region.

Seeds: Let $S_P = \{s_1, s_2, \ldots, s_M\}$ be a set of M potential locations at which seed node infrastructure could be installed, and let *C(s)* denote the cost of deploying a seed node at location $s \in S_P$.

Define a *visibility graph* $G = (V, E)$ with vertex set $V = U \cup S_P \cup \Gamma \cup g^*$, where g^* is a vertex connected by an edge to each POP node in Γ. For each pair $v_i, v_j \in U \cup S_P \cup \Gamma$, G contains an edge $v_i v_j$ if and only if nodes v_i, v_j are within mutual transmission range. The vertex g^* is added to allow us to treat all POP vertices as a single entity.

A potential solution to the seed node placement problem is a subset $S \subseteq S_P$ at which seed nodes are to be deployed. To evaluate the performance of such a selection, we first differentiate between the concepts of *subscription* and *coverage*. It is clearly desirable to a network operator that all user nodes who wish to subscribe at a given time are able to connect to a POP by a sequence of hops. However, if the network is to have the potential for future growth, to be resilient to churn, and have good performance irrespective of the actual locations of subscribers, we must also consider coverage among all potential subscribers. To this end, we say a user $u \in U$ is subscribed if they wish to take part in the network, and we refer to a set of subscribed users $T \subseteq U$ as a *snapshot*. Given a snapshot T and a set of deployed seeds $S \subseteq S_P$, a user or seed $v \;\square\; U * S$ is said to be covered if there exists a path $v, v_{i_1}, v_{i_2}, \ldots, v_{i_k}, g$ of nodes in G, where $v_{i_1}, v_{i_2}, \ldots, v_{i_k} \in T \cup S$ and $g \in \Gamma$. That is, a user is covered if they are connected to a POP by a path of nodes where each is either a deployed seed or another user who also wishes to subscribe. As such, covered users have the potential to communicate with a POP (via a sequence of hops), and thus could successfully join the network if they wished to subscribe. Define the coverage of the snapshot to be the proportion of users in U that are covered. To enable the same terminology to be used for all nodes, we say a seed node $s \in S_P$ is subscribed if and only if it is selected to be deployed (i.e., $s \in S$). We use the term active to describe a node that is both subscribed and covered. All POPs in Γ are subscribed, covered, and active by definition.

Since the coverage of a set of seed nodes is highly dependent on the individual snapshot of users who wish to subscribe, of more interest is the *expected coverage* of the network. From an operators perspective it is important that networks are robust to any variation in the set of subscribers. Let $p_G(v, S)$ denote the probability that node v is covered in the graph G when the set of seeds S is deployed. The expected coverage of S is then

$$E_C(S) = \frac{1}{N} \sum_{u \in U} p_G(v,S) \tag{1}$$

Since we know the probability of subscription for each user, it is possible to derive a recursive procedure to calculate $p_G(v,S)$. First, note that a node is trivially covered if they are within range of a POP $g \in \Gamma$. Otherwise, the probability that a node $v \square U * S$ is covered is the probability that they are within range of an active node (either a seed or user), whose coverage does not depend on the subscription of v. That is

$$p_G(v,S) = \begin{cases} 1 & \text{if } \Gamma \cap N_G(v) \neq \emptyset \\ P\begin{pmatrix} \exists u \in N_G(v): \\ u \text{ is active in } G - v \end{pmatrix} & \text{otherwise} \end{cases} \tag{2}$$

where $N_G(v) = \{w \in V{:}vw \in E(G)\}$ denotes the neighborhood of v in the graph G. The recursive relationship follows by noting that

$$P(u \text{ is active in } G - v) = p_s(u) . p_{G-v}(u,S) \tag{3}$$

However, due to the depth and complexity of recursion necessary, it is not practical to determine $E_C(S)$ exactly for an arbitrary set of seeds S. Instead, we can approximate $E_C(S)$ by simulating the subscription of users over a number of randomly generated snapshots. Given a set of deployed seeds S, we generate a snapshot T of users that wish to subscribe by randomly simulating the decisions of each user u based on their probability of subscription $p_s(u)$. We uniformly generate a random number *rn* □ *[€0,1)* for each user node $u \square U$ and if $p_s(u) > rn$ then u is added to the set T. The coverage *c(S)* can then easily be calculated by examining components in the graph induced[3] by the vertex set $T \cup S \cup \Gamma \cup g^*$ since each node in the same component as g^* must be able to connect to at least one POP via a sequence of hops. Denote the mean value of *c(S)* over a number of snapshots as $\bar{c}$. By considering a sufficiently large set of random snapshots, $\bar{c}$ will approximate E_C.

3 Optimization

Our selection of seed nodes has two objectives: to maximize the expected coverage (equivalently $\bar{c}$) and to minimize the cost of seed nodes deployed. Denote these functions as $f_C(S)$ and $f_S(S)$, respectively. Since $f_C(S)$ and $f_S(S)$ are in direct competition (adding seed nodes increases coverage but with increased expense), it is

[3] The subgraph of G induced by vertex set X has edge set $\{uv{:}uv \in E(G), u,v \in X\}$.

hard to establish a single "best" solution with respect to both objectives simultaneously. The following definitions are useful in addressing this.

Definition 1 (Domination) Let $o_1, o_2, \ldots, o_n$ be objective functions which are to be maximized and let *S* be the set of all possible solutions. $s \square S$ is *dominated* by $t \square S$ (denoted $t \succ s$) if $\square j, j \square \{1, \ldots, n\}$, such that $o_j(t) > o_j(s)$ and $\square i, 1 \leq i \leq n$, (t) $o_j(t) \geq o_j(s)$.

Thus, if solution *s* is dominated by solution *t*, it is always preferential to choose solution *t* over solution *s*.

Definition 2 (Non-dominated set) Given a set *X* of solutions, $x \square X$ is *non-dominated* in X if and only if there is no $x' \in X$ such that $x' \succ x$. The set $\{x \square X : x$ is non-dominated in $X\}$ is called the *non-dominated set* of X.

Thus no solutions from the non-dominated set can be excluded without further knowledge about the relative importance of the objectives.

Definition 3 (Pareto front) Given the set *S* of all possible solutions, the *Pareto front* is the non-dominated set of *S*.

In other words, for any solution within the Pareto front, it is not possible to improve any single objective without a corresponding degradation in one or more of the other objectives. Thus without additional information that ranks the relative importance of objectives, it is impossible to select a single best solution.

Given a set S_P of potential seed node locations, there are $2^{|S_P|}$ possible solutions to the problem; hence for non-trivial problems it is computationally impractical to apply an exhaustive algorithm to find the Pareto front. *Metaheuristic algorithms* that find *near-optimal* solutions in a reasonable time have been successfully utilized for many other design problems in communication network design and configuration. For example, the channel assignment and site selection problem has been approached using algorithms such as *simulated annealing* and *tabu search*. These algorithms intelligently explore the search space, evaluating many possible solutions that improve in quality as the algorithm progresses. Of particular interest are *genetic algorithms*, which mimic the evolutionary process, by maintaining a population of solutions. At each generation, genetic material from pairs of solutions is combined to produce new solutions, named *offspring*, to form the population of the subsequent generation. As a population of solutions evolves, their overall quality should increase from generation to generation. *Mutation* of solutions in the population ensures diversity is introduced. Although these algorithms are general purpose and can be applied to many problems, to be successful, careful consideration must be given to the model and evaluation of potential solutions, particularly where multiple objectives are concerned. Genetic approaches are particularly well suited to multiple objective optimization problems since they maintain multiple solutions that are better able to approximate the Pareto front in comparison to a single solution.

3.1 NSGA-II

The purpose of the optimization is to find a set of mutually non-dominated solutions with respect to $f_C(S)$ and $f_S(S)$ that approximate the Pareto front. To perform the optimization we have implemented the Non-dominated Sorting Genetic Algorithm version II (NSGA-II), introduced in [10]. Here we only give a brief overview of NSGA-II as applied to the seed placement problem in this chapter. See [10] for full details of the general algorithm.

Each solution is represented by a chromosome of M binary values $b_1 b_2 \ldots b_M$, where b_i represents the state of the ith potential seed location. Let $b_i = 1$ denote that a seed node is to be deployed at s_i, and $b_i = 0$ denote that no seed is to be deployed at s_i. There are two solutions that provide the optimum for a single objective. Coverage is clearly maximized if all seeds are deployed, while the infrastructure cost is trivially minimized when no seeds are deployed. Thus we create an initial population P_0 of n solutions consisting of $n - 2$ random chromosomes together with the chromosomes *00...0* and *11...1*.

At each generation t, NSGA-II forms a child population C_t from P_t. Pairs of parents in P_t are selected via binary tournaments. Crossover and mutation are applied to produce a single offspring to be added to C_t from each pair of parents. Uniform crossover is applied, whereby each gene of the offspring is selected from a random parent. Mutation changes the state of each gene with a probability p_m.

Once n offspring have been added to C_t, the parent population P_{t+1} for the next generation is selected from $P_t \cup C_t$. Selection is based on domination and the *crowding distance* of solutions in $P_t \cup C_t$ to ensure good coverage of each objective function.

3.2 Objective Functions

The objective function $f_S(S)$ represents the infrastructure cost and is simply defined as

$$f_S(S) = \sum_{s \in S} C(s) \tag{4}$$

The expected coverage needs to be calculated frequently during the optimization to assess the fitness of each member of the population at each generation. Calculating $\bar{c}$ using simulation is impractical, as it would significantly reduce the number of generations that could be performed. Instead we define $f_C(S)$ using a *surrogate* measure of coverage, which needs to be quick to compute and whose maximization should infer an increase in $\bar{c}$. Here we define $f_C(S)$ based on shortest paths in the subgraph of G induced by the users, POPs, and selected seeds. Define $G(S)$ to be the subgraph of G induced by $S \cup U \cup \Gamma \cup g^*$ and denote by $d_S(v)$ the length of the shortest path in $G(S)$ from v to g^*. Thus $d_S(v) - 1$ denotes the minimum number of hops needed for node v to communicate with a POP. For $v \square S * U$, we define an approximation $p_c^*(v,S)$ to $p_c(v,S)$ by the recursive relationship:

$$p_c^*(v,S) = \begin{cases} 1 & \text{if } \Gamma \cap N_{G(S)}(v) \neq \emptyset \\ 1 - \prod_{w \in N'_{G(S)}(v)} \left(1 - p_s(w) . p_c^*(w,S)\right) & \text{otherwise} \end{cases} \quad (5)$$

where $N'_{G(S)}(v) = \{w \in N_{G(S)}(v) : d_S(w) = d_S(v) - 1\}$. We now calculate the surrogate measure for coverage as

$$f_C(S) = \frac{1}{N} \sum_{u \in U} p_c^*(u,S) \quad (6)$$

The approximation $p_C^*(v,S)$ reduces the complexity of calculation in two respects. First, it only considers paths from v to g^* with the smallest possible number of hops. Second, it treats the probability of these paths being fully active as independent events, where in practice they are dependent. This avoids the need to calculate complex conditional probabilities.

4 Experimental Results

In this section we apply the NSGA-II algorithm described earlier to some random data sets. The aim of the experiments is twofold:

- to demonstrate that the model, objective functions, and optimization algorithms is effective in producing good quality seed node deployments, verified by simulation of user subscriptions;
- to investigate the effect of subscription probabilities on the trade-off between subscription probabilities and number of seed nodes deployed, highlighting key issues that potential network operators must take account of.

4.1 Data Sets

Results are presented for three randomly generated datasets, each constructed with the same parameters. Each data set consists of a 10 by 10 region with a single POP located at random within the region. Hundred user nodes are uniformly randomly distributed at real-valued locations across the region, as are 100 potential seed node locations. Figure 1 shows the first data set used as an example. A uniform propagation environment is assumed and all nodes are assumed to have identical equipment, hence a pair of nodes can form a link provided they are within a communication range r of each other. We assume all potential seeds have equal cost and (unless otherwise stated) that all users have equal probability of subscription.

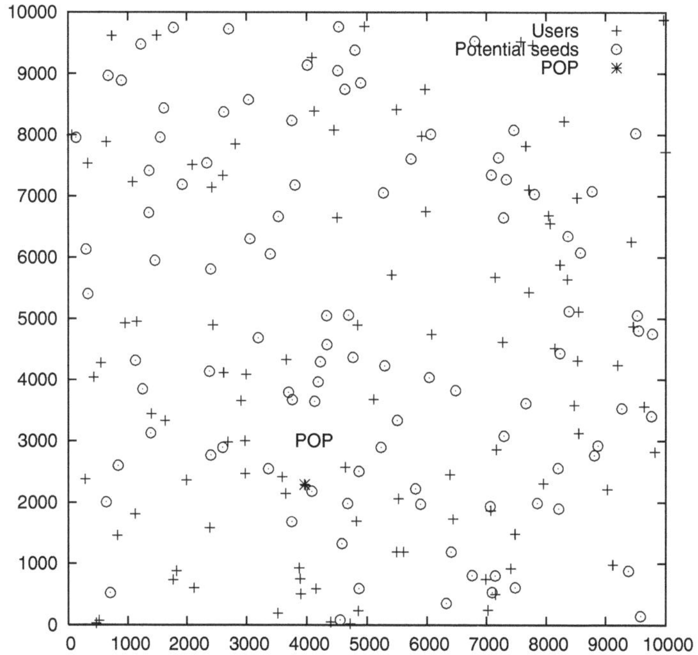

Fig. 1 Data set 1

4.2 Results

The experiments performed in this section use the NSGA-II algorithm to optimize the infrastructure cost measure $f_S(S)$ and the surrogate coverage measure $f_C(S)$. However, all coverage results presented (unless stated otherwise) report the value of $\bar{c}$ generated from the simulation of user subscription over 1,000 snapshots (that is, the surrogate measure of coverage is not reported).

4.2.1 Parameter Tuning

The algorithm has three parameters (number of generations, population size (n), and mutation rate (p_M)) that can be varied, and each has an effect on the performance of the solutions in the final generation. Figures 2, 3, and 4 show the results of varying the population size (shown as the number of random chromosomes in the initial population) and mutation rate on each data set with $p_S(u) = 0.15 \forall u \in U$ and $r = 1.5$ after 1,500 generations. Each data series shows the non-dominated set of the final generation produced by the NSGA-II algorithm. These figures show that higher population size leads to better performance, while mutation rates of 0.1 or 0.2 are clearly preferable to 0 or 1. Since the population size has a direct impact on the computational time, the remaining experiments all use a population of 30 (made up of 28

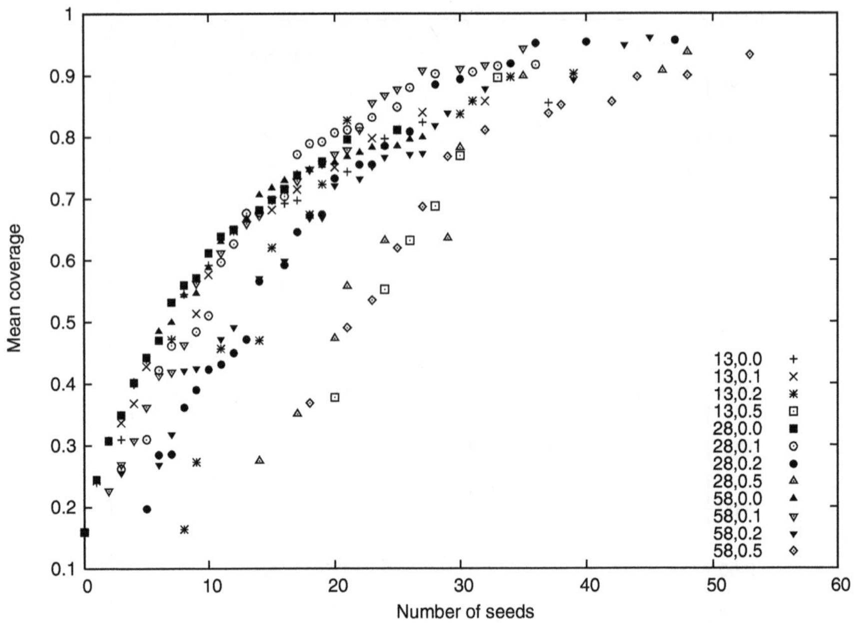

Fig. 2 Variation due to *population size and mutation rate* for data set 1

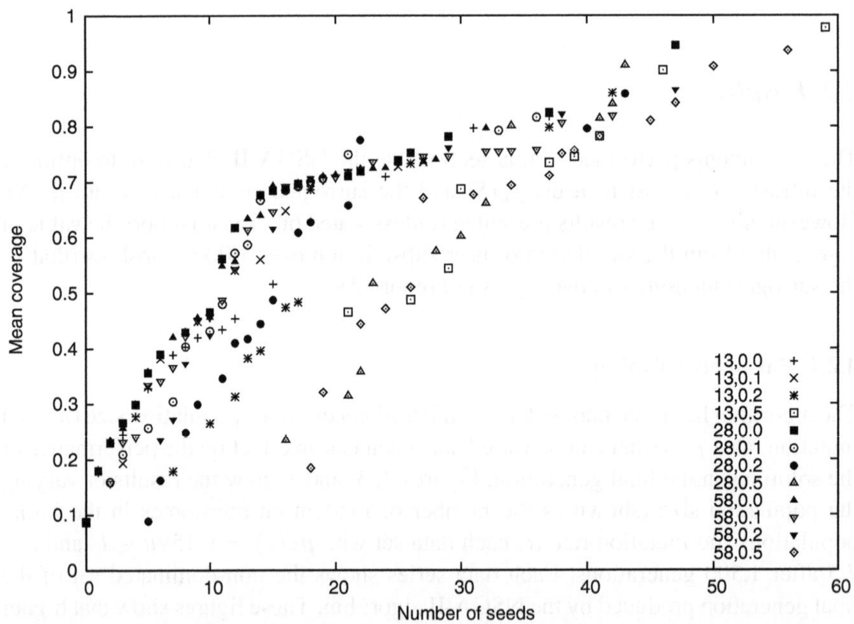

Fig. 3 Variation due to *population size and mutation rate* for data set 2

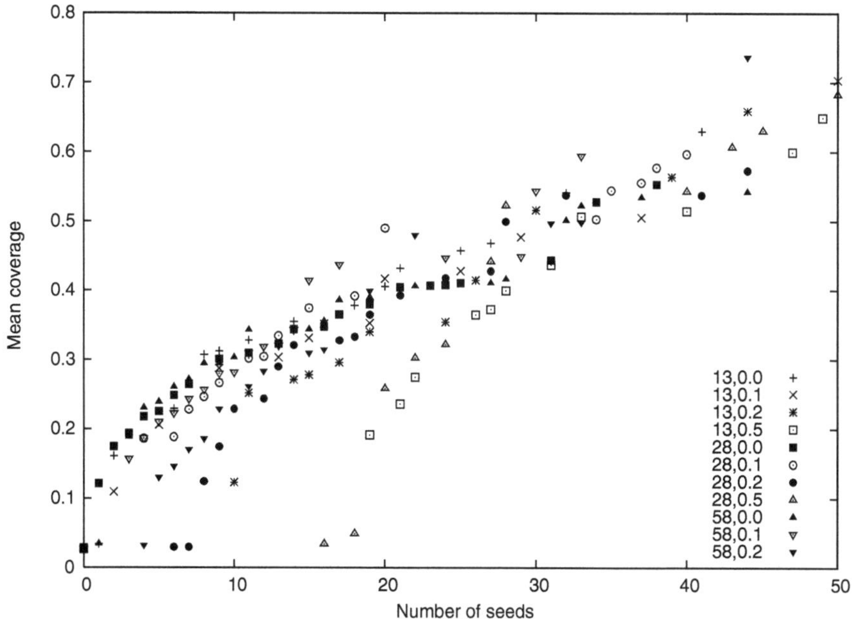

Fig. 4 Variation due to *population size, mutation rate* for data set 3

random chromosomes together with *00...0* and *11...1*), with $p_M = 0.1$ as a compromise between run time and solution quality.

Figure 5 shows the progress of the performance measures[4] for the non-dominated sets during optimization for $p_S(u) = 0.3\ \forall u \in U$, $r = 1.5$, $n = 30$ and $p_M = 0.1$, demonstrating that 1,500 generations are sufficient for the optimization to converge to a stable population. Under these parameters, a single run of the optimization algorithm takes approximately 30 min on a 2 GHz G5 Apple computer.

Figure 6 illustrates the relationship between the surrogate measure $f_C(S)$ and the simulated coverage $\bar{c}$ for the final generation of an optimization run for each data set (with parameters $p_S(u) = 0.3 \forall u \in U$, $r = 1.5$, $n = 30$ and $p_M = 0.1$). Note that although the absolute values are not in close agreement (with the surrogate measure always underestimating the coverage), the relative ordering of the solutions is preserved. Thus, the surrogate is suitable for use in optimizing $\bar{c}$ since increasing $f_C(S)$ implies a similar increase in the simulated coverage results.

4.2.2 Subscription probability

Figures 7, 8, and 9 show the effect of the subscription probability $p_S(u)$ on the trade-off between coverage and infrastructure cost and were generated with $r = 1.5$,

[4] Here the surrogate coverage measure is reported.

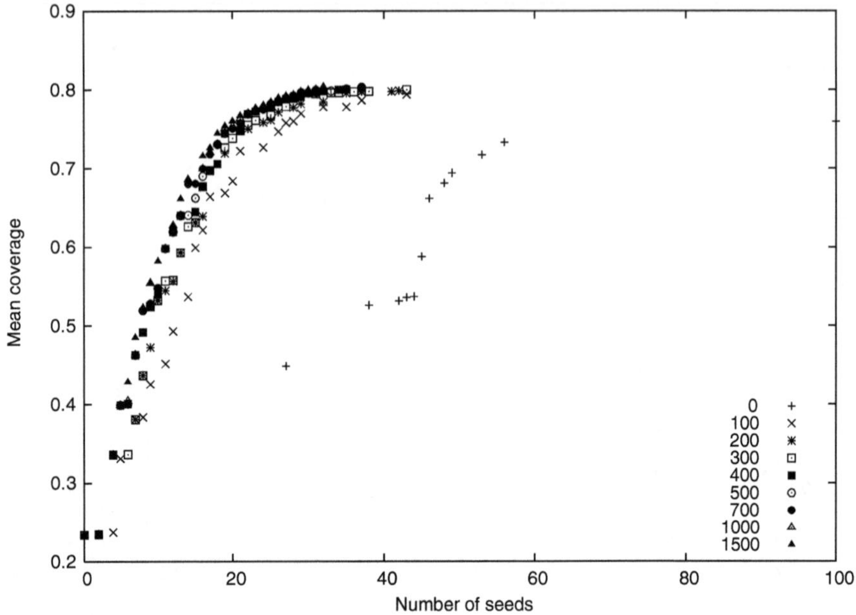

Fig. 5 Progress over generations

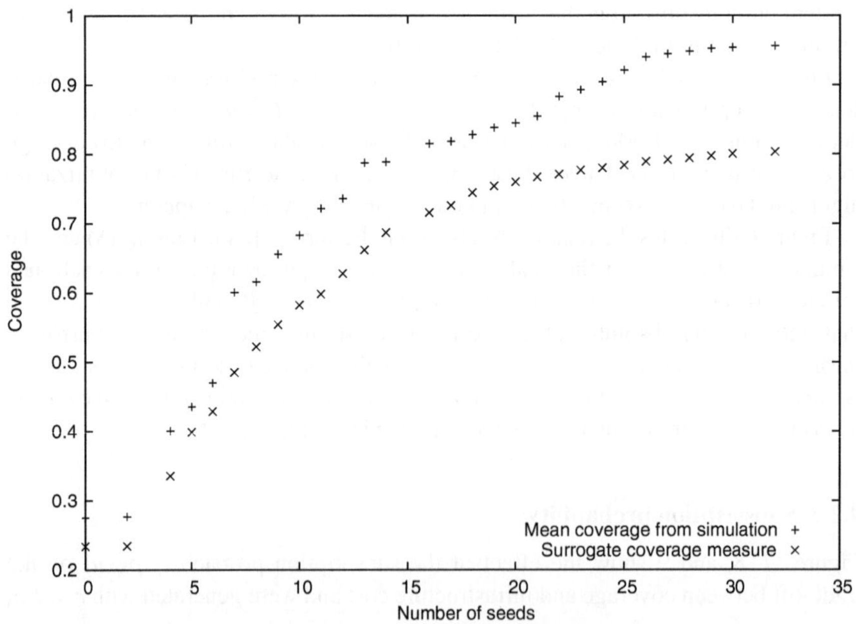

Fig. 6 Correlation between coverage surrogate and simulation for data set 1

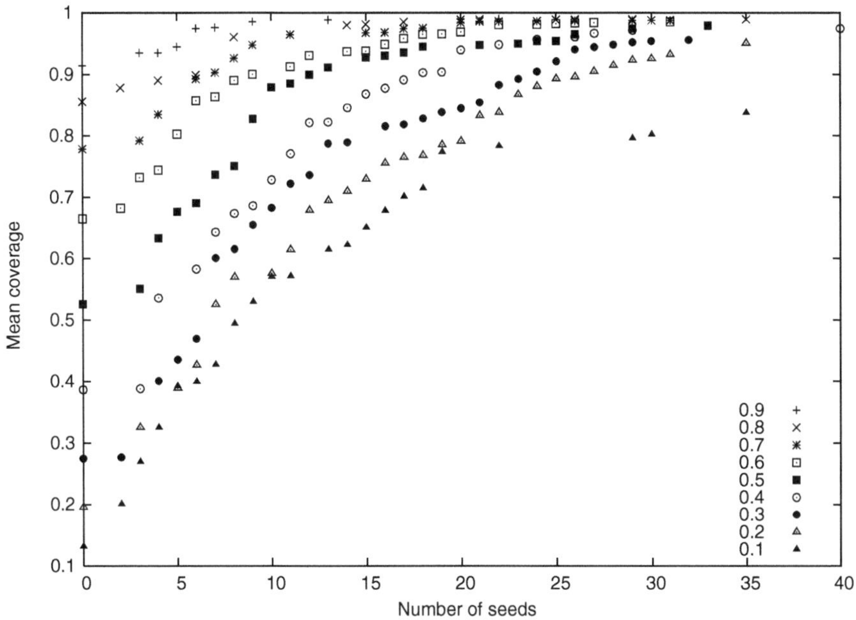

Fig. 7 Effects of *subscription probability* for data set 1

$n = 30$, and $p_M = 0.1$ in all cases. Despite all three having the same density of potential subscribers, there is a marked difference in performance, with coverage consistently higher in data set 1 with fewer seed nodes deployed. It is important to note that the improvement in coverage from each additional seed added decreases rapidly. For all subscription probabilities it is impractical to aim for full coverage due to prohibitively high seed node requirements. This also demonstrates that it is important that a service operator's strategy for improving coverage takes into account the subscription probability. For example, the figures show many places where an increase of 10% in the subscription probability gives an equivalent increase in coverage to adding five seed nodes. It may therefore be more cost effective to reduce subscription costs or increase marketing spend in order to attract further customers rather than simply deploying more infrastructure.

Figures 10, 11 and 12 present an alternative view of the expected network performance, showing the mean proportion of dissatisfied users under the simulation of subscription. That is, the number of nodes that wish to subscribe but are unable to since they are not covered. Again the lack of consistency between the three data sets is clear to see. In particular, data set 3 shows impractical proportion of dissatisfied users at low subscription levels, irrespective of the number of seeds deployed. Furthermore these results show the importance to a network operator of retaining customers, since a small reduction in the subscriber base will require the immediate addition of further seed nodes if existing subscribers are not to be dropped from the network.

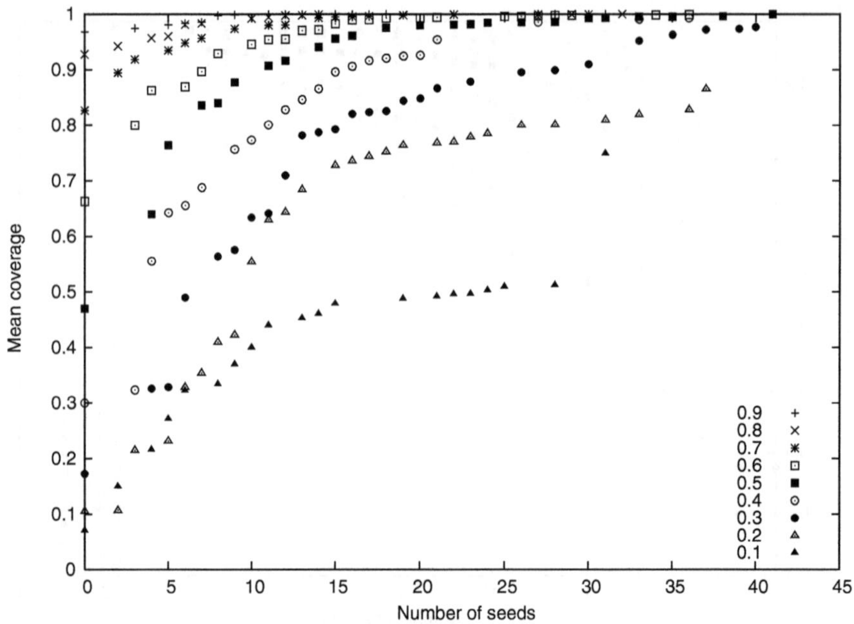

Fig. 8 Effects of subscription probability for data set 2

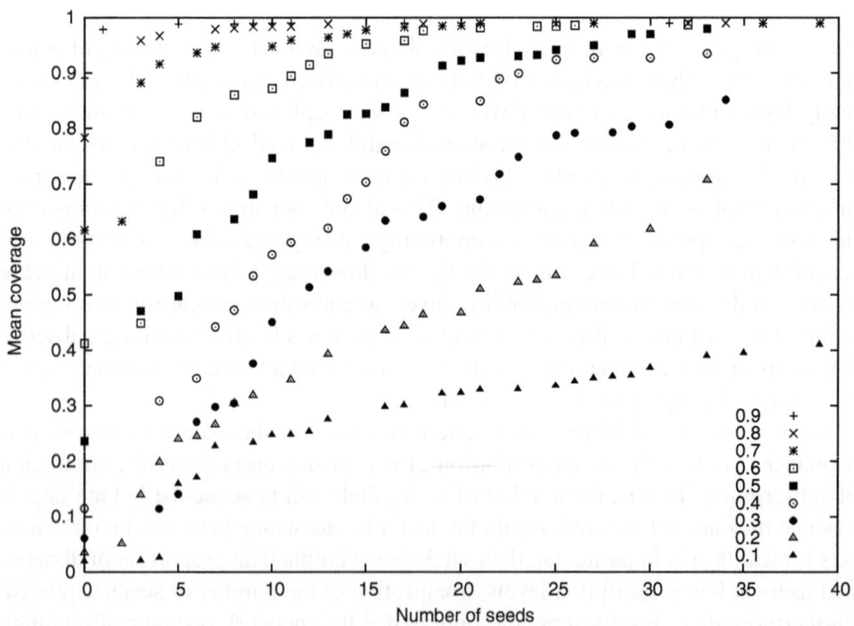

Fig. 9 Effects of subscription probability for data set 3

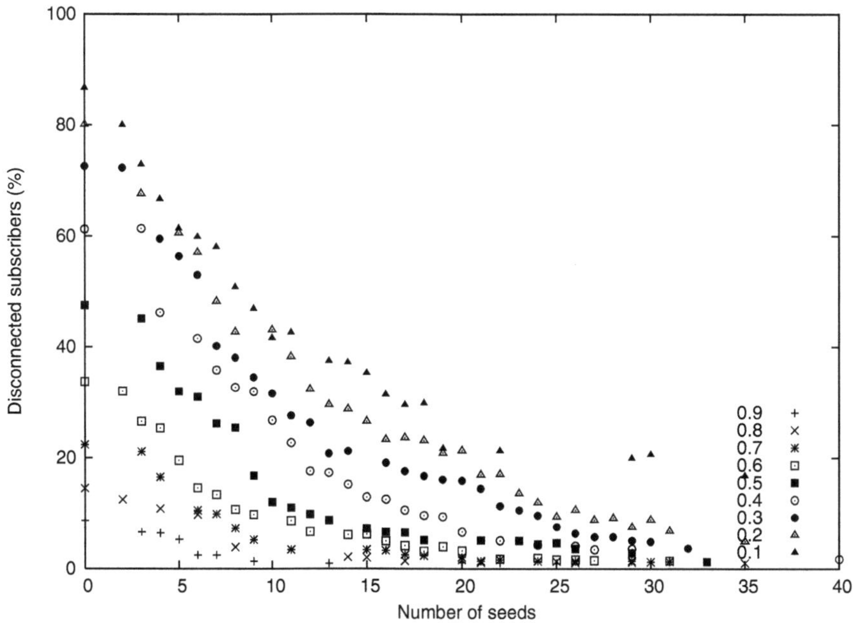

Fig. 10 Dissatisfied subscribers for data set 1 based on *subscription probability*

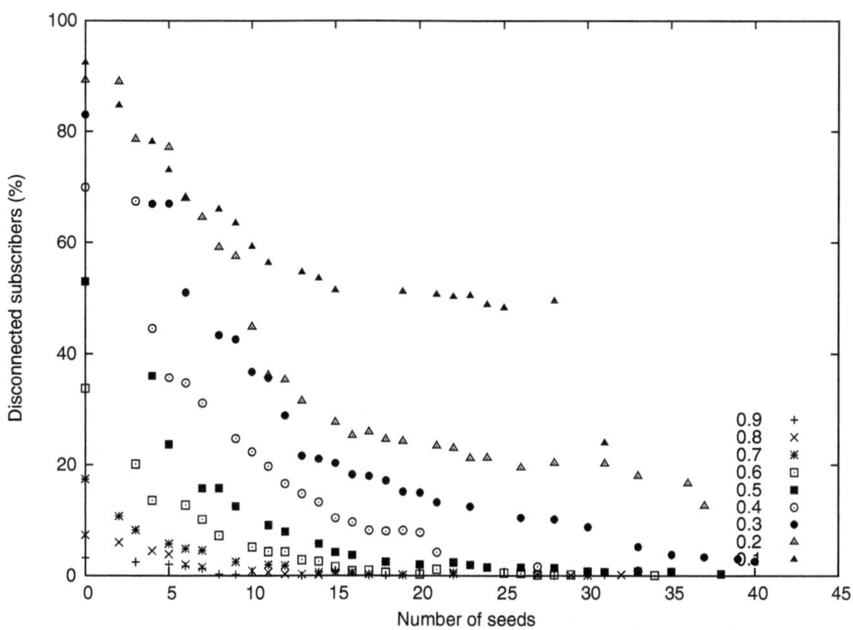

Fig. 11 Dissatisfied subscribers for data set 2 based on *subscription probability*

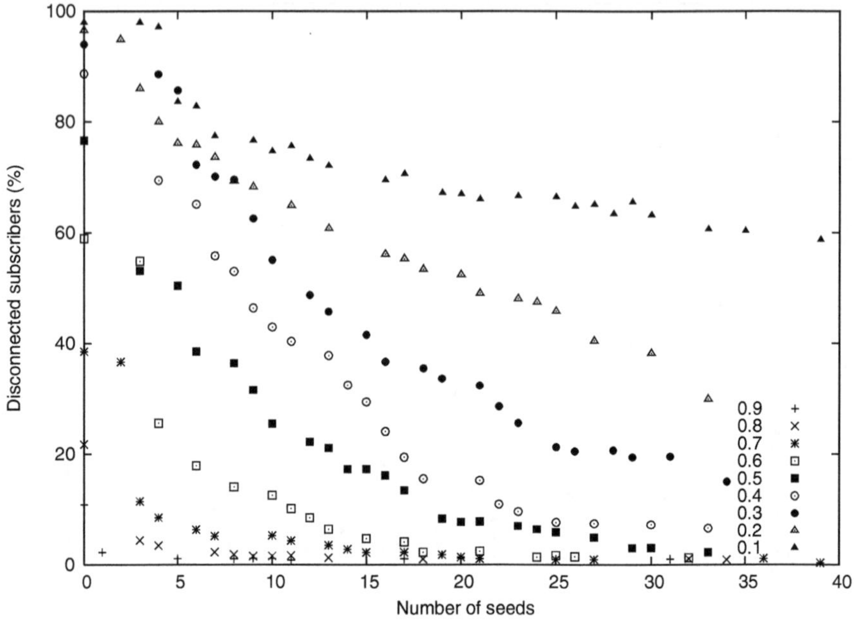

Fig. 12 Dissatisfied subscribers for data set 3 based on subscription probability

Figure 13 demonstrates a secondary benefit of adding more seed nodes, by showing the mean, standard deviation, minimum, and maximum coverage values over the 1,000 simulated snapshots (for the final non-dominated set generated with $p_S(u) = 0.3 \forall u \in U$, $n = 30$, $p_M = 0.1$ and $r = 1.5$). It is clear that additional seeds produce networks that are less dependent on the individual subscription of users, with reduced standard deviation around the mean coverage.

Finally, we note that the seed nodes deployed do not necessarily form a connected backbone to the network on their own, but may rely on the subscription of users to complete the network. Figure 14 demonstrates this for a sample optimization (data set 1, $p_S(u) = 0.3 \forall u \in U$, $r = 1.5$, $n = 30$ and $p_M = 0.1$). This illustrates that there is an associated network planning problem in deploying seed nodes as to when they are installed (should they be deployed from day 1 and risk being redundant or react to subscriptions within the network), and whether they may be removed once the network has achieved a certain level of subscription. For example, it can be seen from Fig. 7 that five seeds are necessary to reach 50% coverage if $p_S(u) = 0.4 \forall u \in U$, however, if the subscription rate subsequently rises to 0.5, the same level of coverage could be achieved if those seed nodes were removed.

5 Conclusions and Future Directions

Infrastructure WMNs have the potential to offer cost-effective last mile access to broadband services to a wide market. However, despite their flexibility and their

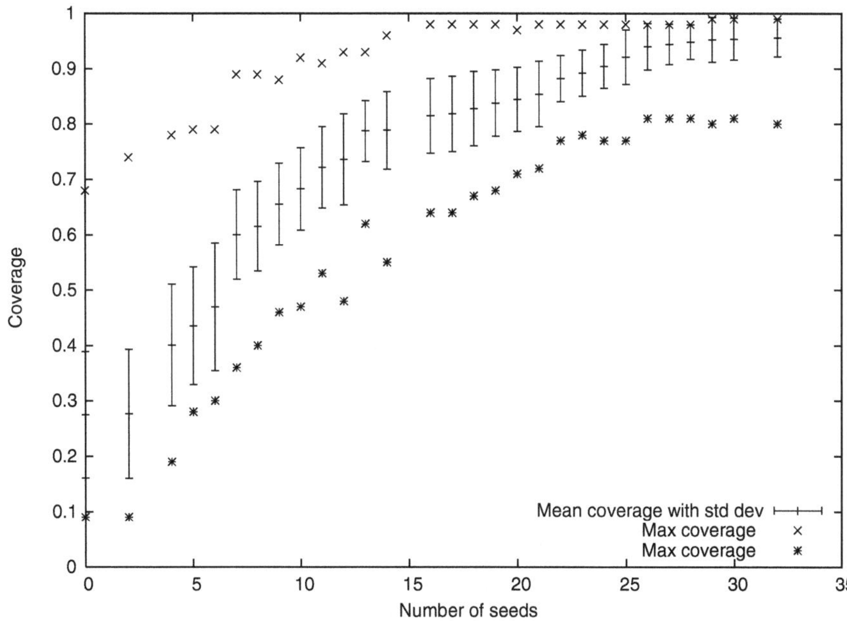

Fig. 13 Variance of coverage in simulation

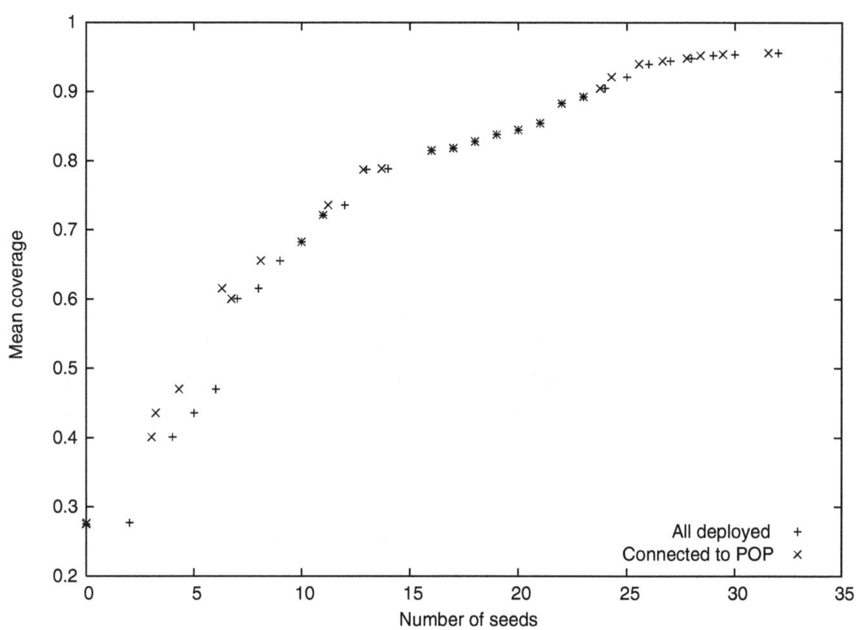

Fig. 14 Connectivity of seed nodes alone

limited need for centralized infrastructure, there are still significant deployment and network management challenges that are frequently overlooked. As an initial step toward some of these, this chapter has presented a model for assessing the coverage of infrastructure wireless mesh networks based on the simulation of consumer subscription decisions. A multiple objective optimization algorithm has been demonstrated that selects near-optimal seed node locations considering the competing objectives of maximizing the expected coverage while minimizing infrastructure cost. Since simulation is too time consuming to be used to evaluate solutions during optimization, a computationally efficient metric for assessing the coverage under uncertain subscription without simulation has been presented. This measure is shown to preserve the order of potential solutions, permitting mathematical optimization techniques to be used successfully. As well as allowing the deployment of individual networks to be undertaken, this represents a promising starting point for studying the key issues in the practical deployment of such networks, as demonstrated by the experiments performed.

Results of simulation studies demonstrate the importance of considering the rate of subscription in evaluating wireless mesh performance. Coverage is poor at low rates, requiring significant levels of investment in infrastructure. The evaluation and optimization techniques proposed allow the trade-off between infrastructure cost and coverage to be determined, and so find the most cost-effective strategy for improving coverage. The results highlight the importance of the take up rate for the service among potential subscribers. There is significant variation in performance depending on the actual subscriber base achieved, suggesting that WMN will require careful management during operation to avoid disconnecting existing subscribers. If coverage is partially dependent on subscribers alone, then operators must be able to rapidly deploy seed nodes to react to customers leaving the service, otherwise large coverage holes will be created. As such, future work should focus on algorithms for dynamic management that can predict in advance where new seeds may be needed, subscribers that are likely to leave and to identify those subscribers that are important for the network performance. This will require modeling of the *expected* quality of service and capacity offered to users. Due to the importance of the subscription probability, understanding the impact of throughput, quality of service, and availability on subscriber behavior will be key. Future work will focus on including the impact of interference and congestion in the expected network performance, for example, to reduce bottlenecks as described in [6]. In this case, seed nodes may also be necessary to improve performance by reducing the number of hops between user and POP. It may also be more cost-effective to deploy further POPs rather than seed nodes, and the model and algorithm can be easily extended to consider this. The probabilistic techniques presented here should also be extended to other uncertainties in network planning, such as in where links may be formed due to obstruction and other propagation effects, which can also be incorporated.

References

1. R. Bruno, M. Conti, and E. Gregori. Mesh networks: commodity multihop ad hoc networks. Communications Magazine, IEEE, 43(3):123–131, March 2005.
2. I.F. Akyildiz, X. Wang, and W. Wang. Wireless mesh networks: a survey. Computer Networks, 47(4):445–487, March 2005.
3. T. Fowler. Mesh networks for broadband access. IEE Review, 47(1):17–22, January 2001.
4. J. Bicket, D. Aguayo, S. Biswas, and R. Morris. Architecture and evaluation of an unplanned 802.11b mesh network. In MobiCom '05: proceedings of the 11th annual international conference on Mobile computing and networking, pages 31–42, New York, NY, USA, 2005. ACM Press.
5. S. Naghian and J. Tervonen. Semi-infrastructured mobile ad-hoc mesh networking. In Personal, Indoor and Mobile Radio Communications, 2003. PIMRC 2003. 14th IEEE Proceedings on, volume 2, pages 1069–1073, China, September 2003.
6. J. Jun and M.L. Sichitiu. The nominal capacity of wireless mesh networks. Wireless Communications, IEEE, 10(5):8–14, October 2003.
7. R. Chandra, L. Qiu, K. Jain, and M. Mahdian. Optimizing the placement of internet TAPs in wireless neighborhood networks. In 12th IEEE International Conference on Network Protocols, volume 00, pages 271–282, Los Alamitos, CA, USA, 2004. IEEE Computer Society.
8. H. Viswanathan and S. Mukherjee. Throughput-range tradeoff of wireless mesh backhaul networks. Selected Areas in Communications, IEEE Journal on, 24(3):593–602, 2006.
9. A. Raniwala, K. Gopalan, and T. Chiueh. Centralized channel assignment and routing algorithms for multi-channel wireless mesh networks. Mobile Computing and Communications Review, 8(2):50–65, 2004.
10. K. Deb, S. Agrawal, A. Pratap, and T. Meyarivan. A fast elitist non-dominated sorting genetic algorithm for multi-objective optimization: NSGA-II. In Lecture Notes in Computer Science, 1917:848–849, 2000.

Use of Non-monotonic Utility in Multi-Attribute Network Selection

Farooq Bari and Victor C.M. Leung

Abstract In the past few decades several wide-area and local-area wireless access technologies have emerged. Network convergence across these different access technologies holds a promise of enabling ubiquitous service availability but faces several technical challenges. With anticipated proliferation of multimode IP devices, the optimal selection of a service delivery network among multiple IP-based wireless access alternatives is one of the important issues that is actively studied and discussed in several standardization forums. Use of multi-attribute decision making (MADM) algorithms has been proposed in the past for network selection decisions in a heterogeneous wireless network environment. A direct comparison of these algorithms is difficult as this would require the use of another MADM algorithm. A better approach instead is to ascertain the appropriateness of the algorithm to the problem space. This chapter provides the basis for evaluating the appropriateness of MADM algorithms for network selection. It analyzes the use of MADM algorithms such as TOPSIS, ELECTRE and GRA for network selection and argues that GRA provides the best approach in scenarios where the utilities of some of the attributes are non-monotonic. We propose a novel stepwise approach for GRA that uses multiple reference networks and explain how it works with network selection scenarios.

1 Introduction

In the past few decades several wide-area and local-area wireless access technologies have emerged. While many of these technologies have been successful in deployment and continue to evolve, none of them provides a universal coverage to cater to the mobile lifestyle of today's user. The desire and expectation to have

F. Bari (✉)
Dept. of Electrical & Computer Engineering, The University of British Columbia, Vancouver, BC, Canada
e-mail: farooq.bari@att.com

S. Powell, J.P. Shim (eds.), *Wireless Technology*, Lecture Notes in Electrical Engineering 44, DOI 10.1007/978-0-387-71787-6_2,

ubiquitous broadband connectivity all the time are the driving forces to push the network operators to look at new and innovative ways of service delivery including the possibility of inter-working of these evolving access technologies with the use of multi-mode devices. An example of multimode device is a terminal that supports Institute of Electrical and Electronics Engineers (IEEE) 802.11 wireless local area network (WLAN) technology and Groupe Special Mobile (GSM) wireless wide-area network (WWAN) technology. A major challenge in this new environment is network selection, i.e., identifying the best-suited service delivery network when the user has multiple networks to choose from. Network selection becomes specially challenging for multi mode devices that have an option to get services from different all IP wireless access types. In the case of current GSM only devices, network selection involves a scan for the network identities, i.e., public land mobile network IDs (PLMN IDs) followed by a selection of one of them based on the pre-provisioned information in the terminal about preferred PLMN IDs and forbidden PLMN IDs. In the case of IEEE 802.11 WLANs, the beacon and probe request/response mechanism provides a way for terminals to discover access points (APs) using service set identifiers (SSIDs). Based on this information and signal strength, the terminal can decide which access point (AP) to associate with. Such simplistic approaches, however, are unlikely to yield optimal network selection in inter-worked heterogeneous all IP broadband systems that use different access technologies while delivering a range of services from a variety of operators. This new environment requires consideration of a number of factors such as the QoS capabilities of the network, current network conditions, QoS requirements of the requested service, and the subscription type of the user.

Network selection is an area of active research and a topic of discussion in several standardization forums. IEEE 802.11u Working Group currently has a draft proposal that would enable information exchange for network selection between the network and the terminal [1]. It leverages the protocol being developed by IEEE 802.21 for this purpose which they also plan to use for selecting networks for vertical handoffs [2]. Similarly 3GPP is looking into the mechanism of network discovery and selection in their work on System Architecture Evolution (SAE) [3]. Along with protocol and architectural aspects of the problem, an essential component in solving the problem of network selection is defining the optimization objective and the algorithm to be used in the selection process. Selection of a non-optimal network creates undesirable results such as poor customer experience or the use of more expensive network. The focus of this chapter is on this aspect of the problem.

1.1 Selection of MADM Algorithm

The requirement to have a consistent service experience for the user requires selection of an optimal delivery network. This issue has special significance for multi-mode IP devices where services can be delivered over a variety of wireless access technologies under varying network conditions. Several factors related to network capabilities and quality of service (QoS) conditions influence the network selection decision process, e.g., bandwidth, delay, jitter, and packet loss. This makes it attractive to use deterministic decision-making tools such as multi-attribute decision making (MADM) algorithms [4]. Their use has been previously considered, e.g., for network selection in a heterogeneous wireless network environment [5–8], to derive a ranking of the available networks in terms of their suitability. The highest ranking network is then selected as the best-suited network. The prior work, however, failed to provide a comparison among the MADM algorithms for use in network selection.

Several alternate MADM algorithms can be suitable for solving a decision problem and the decision maker in this situation can be faced with the task of selecting the most appropriate method from among a number of feasible methods. Classification of MADM algorithms into categories can help to eliminate the algorithms in categories that are not well suited to the problem space, but this process does not provide the most suited algorithm. It is conceivable that a suitable MADM algorithm may be selected for a particular decision problem based on one or both of the following criteria.

1.1.1 Accuracy of the Results Obtained from an Algorithm

For a variety of reasons, different algorithms, when applied to the same problem under the same assumptions, can result in different rankings of the alternatives. In such scenarios it is not possible to objectively rank the MADM algorithms for their ranking accuracy as it would require the use of another MADM algorithm to get such a ranking. For this reason it has been found difficult to use accuracy of the results as a criterion in selecting a specific type of MADM algorithm.

1.1.2 Appropriateness of Applying the Algorithm to the Problem

Because of differences in the approaches used by different MADM algorithms, a direct comparison among them is difficult. It has been proposed in the past that a method which is capable of solving the decision problem and whose decision-making philosophy reflects the values of the decision maker can be considered to be the best suited. Decision makers in general prefer deterministic algorithms that provide reliable results based on a simple and easy to understand philosophy.

So far it has been difficult to perform such an evaluation of the algorithms because of a rather simplistic assumption about the optimization objectives of the decision maker. Prior studies have ignored a possibly diverse range of optimization scenarios based on service and user types that could exist and hence can help in comparing the suitability of the algorithms. In the remainder of this section we describe different QoS profiles that can be used in the decision process which can then lead to a requirement for support of non-monotonic utilities. The following section describes the concept of non-monotonic utilities for the attributes that can help meet a wider variety of optimization objectives. We propose that this concept be leveraged in assessing the suitability of MADM algorithms to network selection.

1.2 Types of QoS Profiles

To better serve their customers, network operators typically use QoS profiles. The selection of a network is highly dependent upon the type of optimization performed for QoS-related attributes stored in such profiles. The two possible QoS profile types that can be stored in user's home network are as follows:

- Overall user QoS profile that is applicable to all of the services that the user is using; e.g., gold, silver, or bronze profile can indicate the level of QoS that the user is expected to have based on the subscription.
- QoS profile of an individual service that is applicable to all subscribers of that service; e.g., VoIP service profile or web-browsing profile.

1.2.1 Service-Based QoS Profile

Based on service-based QoS profile, key service classes can be categorized as follows:

VoIP – This is a low bandwidth application that is very sensitive to delay and jitter but can withstand some packet losses. Transport cost factor is considered negligible because of low bandwidth usage. Also because of low bandwidth requirements, total bandwidth and available bandwidth are not significant factors. Since there is some correlation of utilization with jitter and delay, it is preferred to have a low utilization for the selected network.
Streaming – Being a multimedia service, a streaming application requires a higher bandwidth than VoIP. Therefore available bandwidth, transport cost, and current utilization are important factors. It is less vulnerable to delay and jitter than VoIP because of ability to buffer longer duration of data before play back. Sensitivity to packet loss is similar to VoIP where some packet loss can be compensated without impact to user experience.
Web Browsing – Web-browsing type application is a low QoS service; i.e., the importance of utilization, delay, jitter, and packet loss is low. It does not need a guaranteed bit rate because of spiky nature of web traffic pattern. With statistical traffic multiplexing for such type of traffic, broadband wireless networks can deliver a reasonable customer experience even at lower average data rates. The total bandwidth and allowed bandwidth are therefore less critical but transport cost is considered critical.

1.2.2 Subscription-Based QoS Profile

Based on subscription-based QoS profile the following key subscription classes could be defined.

Gold Subscription – This indicates a premier user subscription that would allow the use of the highest level QoS independent of the transport cost.

Silver Subscription – This indicates a medium priority user subscription that would try to balance between the QoS requirements and other factors such as the transport cost.
Bronze Subscription – This indicates a lower priority user subscription where transport cost is significantly important compared with any QoS parameters.

The examples described in the later section have used service-based QoS profile.

2 Use of Non-monotonic Utilities for Attributes in Network Selection

In general, as part of the decision process, the MADM algorithms associate a measure of suitability or appropriateness, hereafter called utility, with the individual attribute's value. The utility is said to be monotonic if the measure of suitability associated with the attribute shows a monotonic increase or decrease with an increase in attribute value. Otherwise it is said to be non-monotonic. Figure 1 shows a simple decision-making scenario with one attribute, i.e., delay and two networks. The delay attribute is shown with possible monotonic and non-monotonic utilities for different service types. The monotonic utility represents optimization objectives where the network with the least delay value, i.e., Ntwk #1 will be selected for all service types. The non-monotonic utility of delay attribute in Fig. 1 represents the optimization objective of the decision maker where it would like

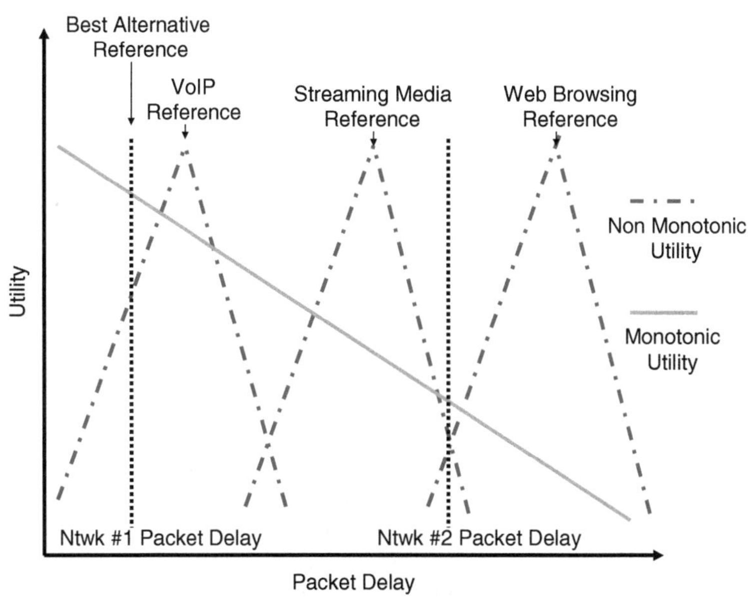

Fig. 1 Graphical representation of a simple decision-making scenario with one attribute and two networks

to use the network closest to the service's delay requirements but not necessarily the best network, which is the network with the least delay. This type of optimization objective would result in selection of Ntwk#1 for VoIP service and use Ntwk#2 for streaming media and web-browsing services. The decision maker may desire this type of optimization for policy reasons such as load balancing across access networks or for keeping the best networks for services and sessions with higher QoS requirements that it can expect to have. It would be similar to the policy of an airline that decide to fly with some first or business class seats empty and not upgrade people from economy class with the knowledge or the hope that it would be able to get full-fare business or first-class customers at the next stop.

In the case described above only one attribute was considered. For the case of multiple attributes, the overall ranking is either obtained via adding the utility associated with each of the attributes or by comparing the utilities for the attributes individually in the decision process. Prior applications of MADM to the network selection problem [3, 6–8] have generally assumed the use of the best network irrespective of service requirements or user type, but the impact of different optimization objectives and hence the use of non-monotonic utility in the decision process have not been considered. This implies a monotonic utility for all attributes. While some of the decision-related attributes can be considered to have monotonically increasing or decreasing utilities, in reality the overall optimization goals of the decision maker may require a combination of monotonic and non-monotonic utilities for different attributes that are taken into considerations during the decision process for network selection. Associating a monotonic increasing or decreasing utility in general with each of the attributes is therefore a simplistic assumption that would limit the scope of types of optimization available to the decision maker [6, 8].

An example of an optimization objective can be to find the network that along with other factors (such as cost) also has the best QoS characteristics from among the list of available networks. In this case the utility of QoS attributes can be considered to be monotonic. However, under a different deployment scenario, the decision maker may wish to assume a non-monotonic utility for some of the attributes considered in the selection process, e.g., as an optimization objective to distribute network traffic across different access networks by selecting the access network offering a QoS closest to that being requested by the service and not the network that may have the best QoS that far exceeds the service's QoS requirements.

This chapter analyzes the suitability of several most commonly used MADM algorithms for the problem of optimal network selection, where not all the attributes considered in the decision-making process have a monotonically increasing or decreasing utility. Such network selection scenarios will be quite common in future heterogeneous wireless network environments used for delivery of both real-time and non-real-time services. The algorithms considered in the analysis are TOPSIS [9], ELECTRE [9, 10], and GRA [11]. Among these MADM algorithms, GRA is found to be the most suited for optimization objectives requiring both monotonic and non-monotonic utilities of attributes. Using a heterogeneous wireless network environment as an example, we demonstrate how the algorithm can be implemented to achieve different optimization objectives and its impacts on the resulting network ranking for network selection.

3 Comparison of MADM Algorithms for Use with Non-monotonic Utilities of Attributes

3.1 TOPSIS

TOPSIS (Technique for Order Preference by Similarity to Ideal Solution) is a widely used MADM algorithm that was developed by Yoon and Hwang [9]. The algorithm calculates perceived positive and negative ideal solutions based on the range of attribute values available for the alternatives and selects the best solution as the one with the shortest distance to the positive ideal solution and longest distance from the negative ideal solution. The distances are measured in Euclidean terms. Because of the concept of positive and negative ideal solutions that use Euclidean distances, a standard implementation of TOPSIS requires that the utilities of the attributes under consideration increases or decreases monotonically.

3.2 ELECTRE

The ELECTRE (Elimination et choix traduisant la realite) method [9, 10] performs pairwise comparisons among all alternatives for each one of the attributes separately in order to develop outranking relationship between the alternatives. In its standard implementation, the method first removes the less desirable alternatives and then using a complimentary analysis it selects the best-suited alternatives. Since the comparison is direct among the available alternatives there is no concept in ELECTRE of comparing the alternatives to some reference set of values to see how close the parameters values are to the desired values. The notion of a monotonically increasing or decreasing utility of an attribute is inherent in direct comparison among the alternatives, which makes standard ELECTRE algorithm not well suited for use with attributes having non-monotonic utilities.

3.3 GRA

GRA (gray relational analysis) is another very popular decision-making technique that is based on gray system theory. Originally developed by Deng [11], gray systems theory has been applied to solve a variety of real-life problems ranging from the fields of business, operations research, and engineering, to social sciences. One of its areas of application has been MADM, decision making where multiple attributes influence the decision process. Unlike other MADM algorithms, GRA uses a reference set of attribute values for comparison with attribute values of the alternatives. It has been applied in the past [6, 7] to solve the problem of network selection in a heterogeneous network environment. The problem of selection of an optimal network in a heterogeneous environment, however, is quite complex and it is possible to apply GRA in several different ways to address the problem. In the prior work, the utility aspects of the algorithm were not explored and a single reference network was constructed, which implied monotonic utilities for all the attributes for all service or user types. Because of its ability to use reference attribute values in the decision process, GRA can be applied where the optimization objectives require non-monotonic utilities for some of the attributes and monotonic utilities for the others. As described in Section 5, such an implementation of GRA would use multiple reference networks. The network rankings in this case could be quite different than if monotonic utilities were considered for all the attributes.

It is clear that for optimization scenarios where a utility does not increase or decrease monotonically with an increase or decrease in attribute value, standard implementations of MADM algorithms such as TOPSIS and ELECTRE will have limited applicability. Other simpler compensating MADM algorithms such as SAW (Simple Additive Weighing) [9] and WPM (Weighed Product Method) [9] also have similar limitations because of their inherent assumption about monotonic utilities of attributes. These MADM algorithms do, however, allow assigning different weights to the attributes before they are combined to calculate ranking indices. This general feature of MADM algorithms can be used to apply algorithms such as TOPSIS to network selection for different service types. For example, services that are less sensitive to QoS and more sensitive to transport cost (e.g., web browsing) could have higher weights assigned to the cost attribute and lower weights assigned to the QoS attribute. This would allow alternatives that are closer to the positive ideal solution in transport cost (i.e., lowest in cost) to get more importance in decision making than network alternatives with QoS-related attribute values closest to the positive ideal solution (i.e., best QoS attribute value). However, it is important to note that this type of applicability would provide a different optimization than trying to find the network alternative which QoS attributes are closest to those of the requested service. As stated earlier, GRA favors a selection that gives a closest match to a set of reference data values. This process inherently supports the notion that these reference values do not necessarily need to be the best or the worst values associated with the attributes. In addition, it also has the ability to assign different weights to different attributes. These two tunable aspects of GRA when combined provide a much better mechanism to achieve optimization objectives involving attributes with non-monotonic utilities. The rest of this chapter describes the application of GRA to the problem of network selection with some attributes having non-monotonic utilities.

4 Theory of Gray Relational Space

Gray Relational Analysis (GRA) has sometimes been compared to fuzzy logic. However, GRA is different from fuzzy logic. While both can help in decision making under uncertain conditions, fuzzy logic deals with imprecise information and GRA deals with insufficient or scarce information. GRA is based on the concept of gray relational space (GRS). GRS (X,Y) describes a relationship Y between reference data values X_0 and sequence of data values X. So if $y \in Y$, $x_i \in X, x_0 \in X_0$ such that $x_0 = x_0(1),\ldots\ldots x_0(n)$ and $x_i = x_i(1),\ldots\ldots x_i(n)$ then $y(x_0(k),x_i(k))$ would represent a GRS at point *k* provided the axioms documented in [11] are satisfied. In addition, a gray relational grade for a series *i* could then be represented as

$$y(x_0,x_i) = \frac{1}{n}\sum_{k=1}^{n} y(x_0(k),x_i(k)).$$

A representation of $y(x_0(k),x_i(k))$ that satisfies all of the axioms in [11] is represented as

$$y(x_o(k),x_i(k)) = \frac{\min_i \min_k |x_0(k) - x_i(k)| + \xi \max_i \max_k |x_0(k) - x_i(k)|}{|x_0(k) - x_i(k)| + \xi \max_i \max_k |x_0(k) - x_i(k)|}$$

where $0 \leq \xi \leq 1$ is called a distinguished coefficient and $y(x_0(k),x_i(k))$ is called the gray relational coefficient (GRC). When applying GRA to ranking of networks while selecting a network, GRC is a measure of how closely a network's attributes match the reference network's attributes. In this respect it represents the overall utility that takes into consideration individual values of all the attributes. The higher the value of GRC, the closer would be the candidate network to the reference network. Hence, for the purposes of network selection GRC is equivalent to a utility that takes into consideration all individual attribute values.

5 Application of GRA Adapted to Network Selection with Non-monotonic Utility

For the network selection problem in this chapter, Table 1 provides a typical set of attributes that can be considered in such a decision-making process.

Using the attributes defined above, a candidate network NW for evaluation by GRA can be represented as follows:

$$NW = \begin{bmatrix} CB\ TB\ AB\ U\ D\ J\ L \end{bmatrix}$$

If there are *N* alternative networks to be considered in the selection process, they can be represented in the form of a matrix as follows:

Table 1 Attributes for network selection

Attribute	Brief Explanation
Cost per Byte CB	Data transport cost on a particular access system
Total Bandwidth TB	Overall bandwidth of the wireless access link
Allowed Bandwidth AB	Bandwidth per user allowed by the access system
Utilization U	Current utilization of the wireless link
Packet delay D	Average packet delay within the access system
Packet Jitter J	Average packet jitter in the access system
Packet Loss L	Average packet loss rate in the access system

$$NW_i = \begin{bmatrix} CB_1 & TB_1 & AB_1 & U_1 & D_1 & J_1 & L_1 \\ CB_2 & TB_2 & AB_2 & U_2 & D_2 & J_2 & L_2 \\ . & . & . & . & . & . & . \\ . & . & . & . & . & . & . \\ . & . & . & . & . & . & . \\ CB_N & TB_N & AB_N & U_N & D_N & J_N & L_N \end{bmatrix}$$

A reference access network is needed for application of GRA. In the case of monotonically increasing or decreasing utilities for the attributes, this reference network can be developed by using the maximum or minimum value of the attributes. In this case there will only be one reference network. However, if there are services that have different QoS requirements (VoIP, streaming, web browsing, etc.), or the users are of different categories (e.g., bronze, silver, gold), then the decision maker can use different reference networks for each one of the categories. A reference network in this case can be created based on the information about the user/terminal preferences, e.g., indication of the requested service, or based on the user profile stored in the home network, e.g., the subscribed QoS. These multiple reference networks would result in non-monotonic utilities for some of the attributes. Table 3 shows four different reference networks developed for the example described in Section 6. The reference network *i* for a particular service or user type can therefore be represented as follows:

$$(NW_{ref})_i = ((CB_{ref})_i \ (TB_{ref})_i \ (AB_{ref})_i \ (U_{ref})_i \ (D_{ref})_i \ (J_{ref})_i \ (L_{ref})_i)$$

The units of measurement for the attributes such as cost, bandwidth, and delay will be different. In order to apply the algorithm without having the artifacts related to different units of measurement impacting the results, the attributes will have to be made unit-less before they can be directly compared or combined during the calculations. This process is called normalization. There are several normalization techniques e.g., dividing attribute value with a maximum value for that attribute across all the alternatives. Using these normalized attribute values, an updated matrix is created as follows:

$$\tilde{NW}_i = \begin{bmatrix} \tilde{CB}_1 & \tilde{TB}_1 & \tilde{AB}_1 & \tilde{U}_1 & \tilde{D}_1 & \tilde{J}_1 & \tilde{L}_1 \\ \tilde{CB}_2 & \tilde{TB}_2 & \tilde{AB}_2 & \tilde{U}_2 & \tilde{D}_2 & \tilde{J}_2 & \tilde{L}_2 \\ \cdot & \cdot & \cdot & \cdot & \cdot & \cdot & \cdot \\ \cdot & \cdot & \cdot & \cdot & \cdot & \cdot & \cdot \\ \cdot & \cdot & \cdot & \cdot & \cdot & \cdot & \cdot \\ \tilde{CB}_N & \tilde{TB}_N & \tilde{AB}_N & \tilde{U}_N & \tilde{D}_N & \tilde{J}_N & \tilde{L}_N \end{bmatrix}$$

The reference network i*'s attributes are also normalized and* a normalized reference network vector is created as follows:

$$(\tilde{NW}_{ref})_i = ((\tilde{CB}_{ref})_i\ (\tilde{TB}_{ref})_i\ (\tilde{AB}_{ref})_i\ (\tilde{U}_{ref})_i\ (\tilde{D}_{ref})_i\ (\tilde{J}_{ref})_i\ (\tilde{L}_{ref})_i)$$

If the reference attribute values lie outside of the attributes values for the alternatives under considerations, then calculation of the maximum and minimum values to be used in the normalization process should include the reference values as well.

Distance vectors are calculated for attributes of each access network under consideration by taking the absolute difference between the attribute of the reference network and the corresponding attribute of the candidate network. For example, in the case of TB for network i, the distance value from reference network j is calculated as follows:

$$(\Delta_{TB})_i = |(\tilde{TB}_{ref})_j - \tilde{TB}_i|$$

The matrix of distance value for each of the attributes for the *N* networks under consideration can therefore be created as follows:

$$(\Delta_{NW})_i = \begin{bmatrix} (\Delta_{CB})_1 & (\Delta_{TB})_1 & (\Delta_{AB})_1 & (\Delta_U)_1 & (\Delta_D)_1 & (\Delta_J)_1 & (\Delta_L)_1 \\ (\Delta_{CB})_2 & (\Delta_{TB})_2 & (\Delta_{AB})_2 & (\Delta_U)_2 & (\Delta_D)_2 & (\Delta_J)_2 & (\Delta_L)_2 \\ \cdot & \cdot & \cdot & \cdot & \cdot & \cdot & \cdot \\ \cdot & \cdot & \cdot & \cdot & \cdot & \cdot & \cdot \\ \cdot & \cdot & \cdot & \cdot & \cdot & \cdot & \cdot \\ (\Delta_{CB})_N & (\Delta_{TB})_N & (\Delta_{AB})_N & (\Delta_U)_N & (\Delta_D)_N & (\Delta_J)_N & (\Delta_L)_N \end{bmatrix}$$

Gray relational coefficient (GRC) is a measure of similarity of an attribute to its reference value. It is calculated for each of the matrix entries. For example, in the case of TB, it is calculated as follows:

$$(GRC_{TB})_i = \frac{\Delta\min + \zeta\Delta\max}{(\Delta_{TB})_i + \zeta\Delta\max}$$

where $\zeta \in [0,1]$ and $\Delta\min$ and Δ maxcan be calculated as follows:

$$\Delta\max = \max_i (\Delta CB_i + \Delta TB_i + \Delta AB_i + \Delta U_i + \Delta D_i + \Delta J_i + \Delta L_i)$$

$$\Delta\min = \min_i (\Delta CB_i + \Delta TB_i + \Delta AB_i + \Delta U_i + +\Delta D_i + \Delta J_i + \Delta L_i)$$

The next step is to consider the relative importance of each of the attributes in the decision about network selection. For this purpose each of the attribute is assigned a weight "w" such that

$$W = W_{TB} + W_{AB} + W_{U} + W_{D} + W_{J} + W_{L} = 1$$

The new weighted GRC matrix will be as follows:

$$GRC = \begin{bmatrix} W_{CB}{}^{*}(GRC_{CB})_1 & W_{TB}{}^{*}(GRC_{TB})_1 & W_{AB}{}^{*}(GRC_{AB})_1 & W_{U}{}^{*}(GRC_{U})_1 & W_{D}{}^{*}(GRC_{D})_1 & W_{J}{}^{*}(GRC_{J})_1 & W_{L}{}^{*}(GRC_{L})_1 \\ W_{CB}{}^{*}(GRC_{CB})_2 & W_{TB}{}^{*}(GRC_{TB})_2 & W_{AB}{}^{*}(GRC_{AB})_2 & W_{U}{}^{*}(GRC_{U})_2 & W_{D}{}^{*}(GRC_{D})_2 & W_{J}{}^{*}(GRC_{J})_2 & W_{L}{}^{*}(GRC_{L})_2 \\ \vdots & \vdots & \vdots & \vdots & \vdots & \vdots & \vdots \\ W_{CB}{}^{*}(GRC_{CB})_N & W_{TB}{}^{*}(GRC_{TB})_N & W_{AB}{}^{*}(GRC_{AB})_N & W_{U}{}^{*}(GRC_{U})_N & W_{D}{}^{*}(GRC_{D})_N & W_{J}{}^{*}(GRC_{J})_N & W_{L}{}^{*}(GRC_{L})_N \end{bmatrix}$$

Using the GRC matrix thus calculated, the gray relational coefficient for each of the candidate network is calculated as follows:

$$(GRC_{NW})_i = W_{CB}{}^{*}(GRC_{CB})_i + W_{TB}{}^{*}(GRC_{TB})_i + W_{AB}{}^{*}(GRC_{AB})_i + W_{U}{}^{*}(GRC_{U})_i + W_{D}{}^{*}(GRC_{D})_i + W_{J}{}^{*}(GRC_{J})_1 + W_{L}{}^{*}(GRC_{L})_i$$

The network with the highest value of gray relational coefficient is considered to be the best network.

6 Evaluation of Using Non-monotonic Utilities in a Heterogeneous Wireless Network Environment

To evaluate the impact of different optimization objectives, we consider a network selection scenario with five networks. For each of these networks, the attribute values to be used in the decision process are shown in Table 2. The table provides the numerical attribute values for these five networks that were used for illustrative purposes in the decision process for network selection. They are representative of listed example network types that a typical user could expect. For example, the cost attribute is derived on the basis of spectral efficiency of the technology and whether the technology runs on licensed or unlicensed spectrum. So the cost is lowest for unlicensed spectrally efficient technology such as IEEE 802.11n and it is highest for licensed and relatively less spectrally efficient technology such as 3G. Also the

Table 2 Attribute values for alternative networks at the time of network selection

	CB (%)	TB (mbps)	AB (mbps)	U (%)	D (ms)	J (ms)	L (per10^6)
Ntwk#1, e.g., 3G#1	100	2	0.2	10	400	50	100
Ntwk#2, e.g., 3G#2	100	2	0.4	5	200	25	50
Ntwk#3, e.g., 802.11a	10	54	2	20	100	15	15
Ntwk#4, e.g., 802.11n	5	100	5	40	150	30	20
Ntwk#5, e.g., 4G	30	100	5	20	100	20	15

maximum estimated throughput for each of the example technologies has been used for the total bandwidth attribute. The allowed bandwidth has been assumed to be based on operator policy to rate limit its customers differently for different access technologies. Other QoS related attributes such as delay, jitter and loss represent a snapshot of the values that could exist in these networks at the time of decision.

In our case we address the network selection problem for three distinct types of services, namely, VoIP, streaming media, and web browsing. Each of these service types has its distinct set of QoS requirements. In the following subsection we describe how to use an adapted version of GRA for the scenarios under consideration.

6.1 Setting up GRA for Network Selection

The GRA algorithm can be applied in more than one ways to the problem of network selection. However, it is well suited to handling diverse optimization objectives including those requiring non-monotonic utility of attributes. Figure 2 shows three different ways of application of the GRA algorithm. We recommend the third approach as described earlier since it provides the maximum flexibility for tuning the algorithm to different optimization objectives. A two-step process for tuning GRA is proposed below.

6.1.1 Determine Reference Attribute Values for Different Service or User Categories

Based on optimization criteria derived from the QoS requirements for the services or user types, reference networks are created by the decision maker (e.g., user's home network). It will be toward these reference values that the GRA will try to find a closest match from a given list of alternative networks. This step relates to addressing the non-monotonic nature of utility for some of the attributes under consideration. For example, the VoIP reference network's attribute values would reflect higher QoS requirements compared to a reference network for web browsing. Creating reference networks is a one-time event and can therefore be provisioned into the decision process.

6.1.2 Determine Attribute Weights for Different Service or User Categories

To allow further tuning of the GRA algorithm to the optimization objectives, the weights (i.e., the importance) assigned to the attributes used in decision making are adjusted for each type of service. The weight assigned reflect the relative importance of an attribute for that service or user type. For example, the cost attribute would carry relatively higher weight for streaming type service when compared with VoIP service. The process of determining attribute weights is also a one-time event and can be provisioned into the decision process.

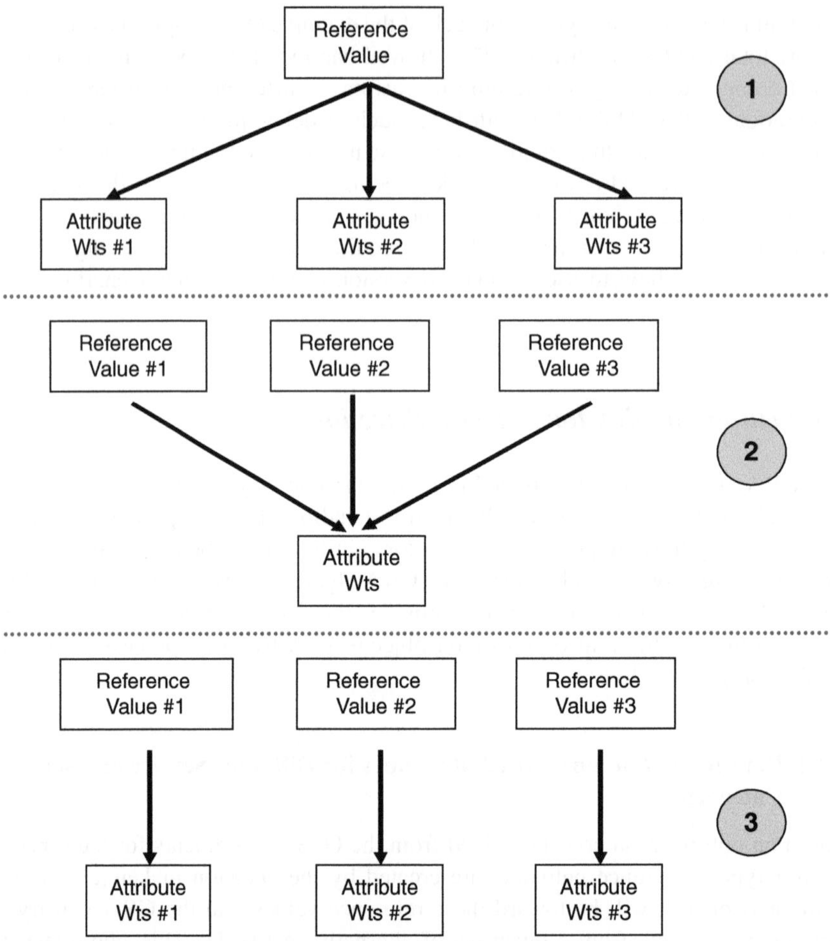

Fig. 2 Three ways of using reference attribute values and attribute weights with GRA algorithm that were considered. Approach 3 is recommended in network selection

Table 3 shows four reference networks that were used in the evaluation. The reference values for the attributes are considered the values toward which the operator would like to optimize while selecting a network from among the alternatives. For example, by adopting the utilization of the lowest utilized alternative network as a reference attribute value, the decision maker could help improve the utilization for under utilized alternative networks and hence have an improved balance of traffic loads across networks. However, this may not be always the optimization criteria and other decision makers may like to have different criteria for selecting this and other attribute values. So the first reference network is created from best values for the attributes from among the alternatives networks. The remaining three reference networks use a combination of best attribute values (for CB, TB, and U) and reference values derived from QoS requirements for the service types (for AB, D,

Table 3 Reference attribute values

	CB (%)	TB (mbps)	AB (mbps)	U (%)	D (ms)	J (ms)	L (per 10^6)
Best QoS	5	100	5	10	100	15	15
VoIP	5	100	0.02	10	100	15	15
Streaming	5	100	1	10	400	50	50
Web browsing	5	100	0.1	10	1,000	100	100

J and L). The entries in Table 3 show that some of the reference values generated from specific service types lie outside the range of attribute values for the network alternatives under consideration. This can potentially cause problems in the normalization process. In order to avoid possible ranking abnormalities, the normalization process is modified by appropriately adjusting the minimum/maximum values used in order to include reference values as well.

The assigned weight for each of the attributes for different service categories considered in the evaluation is shown in Table 4. For example, the importance of transport cost was considered high for web browsing as compared to VoIP. To evaluate the impact of assigned weights, a set of scenarios was also evaluated where only one weight distribution was used for all the different service types.

Scenario 1 uses the best values as the reference for all attributes but uses different weights for different service types. This is shown in the first approach of Fig. 2. Scenario 2 only changes the reference attribute values while keeping the same attribute weights for all service types. This is shown in the second approach of Fig. 2. Scenario 3 calculates ranking when the reference attribute values as well as the attribute weights were changed for different services as shown in approach 3 of Fig. 2. For example, in the case of streaming media type service, it first compares the network attribute values with the reference values specific to streaming media type service to find the degree of match for each of the individual attributes. Then based on the emphasis that should be placed on the degree of match for each of the attributes, a weight is assigned to it as explained earlier.

Table 4 Assignment of attribute weights

Attribute weights used for scenarios 1 and 3

	CB	TB	AB	U	D	J	L
VoIP	0.05	0	0	0.2	0.3	0.3	0.15
Streaming	0.2	0.15	0.2	0.2	0.1	0.1	0.05
Web browsing	0.5	0.05	0.15	0.1	0.05	0.05	0.1

Attribute weights used for scenarios 2

	CB	TB	AB	U	D	J	L
VoIP, streaming, web browsing	0.2	0.15	0.2	0.2	0.1	0.1	0.05

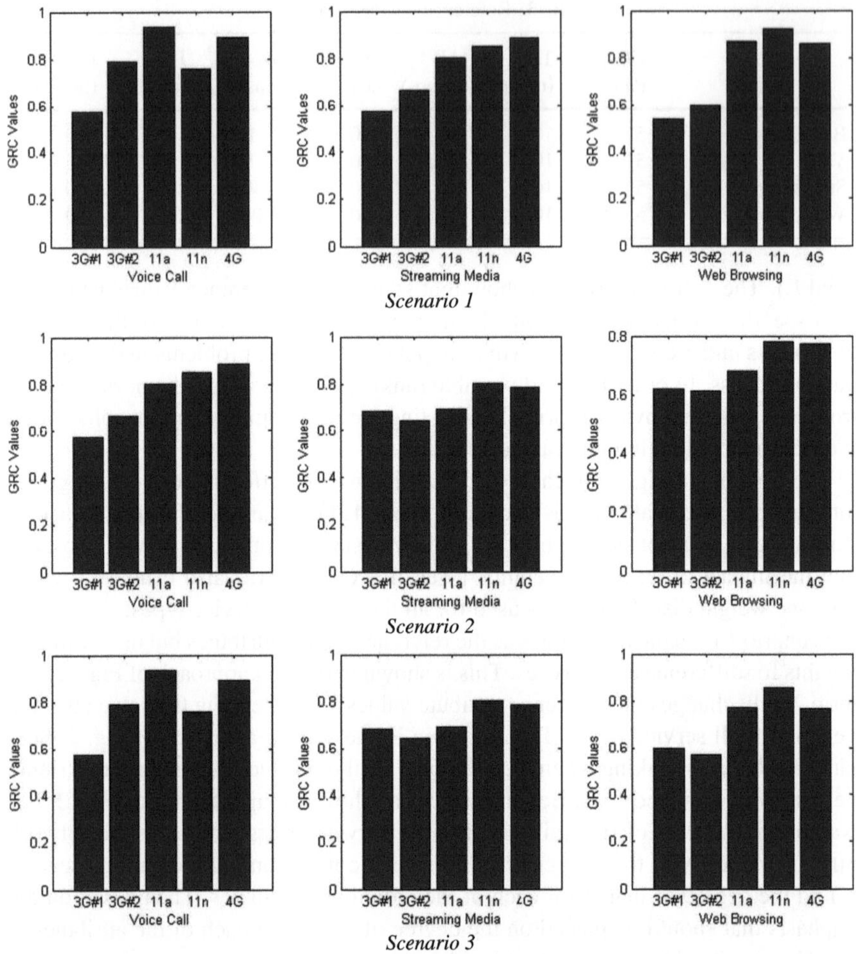

Fig. 3 Results for network selection for three possible configurations of GRA shown in Fig. 2. Configuration 3 is preferred as it provides maximum flexibility to fine tuning the algorithm to optimization objectives

Figure 3 shows the results for all the scenarios that were evaluated and hence shows the impact of use of multiple reference networks and attribute weights. As explained earlier it is not possible to directly evaluate results of MADM algorithm in terms of accuracy. However, since scenario 1 uses a combination of reference values and attribute weights to distinguish among different service types, it will provide a more balanced approach with results closest to optimization objectives of the decision maker if the intent was to select the network closest in characteristics to the reference network for that service type. This is also apparent by comparing results for scenarios 1 and 3, which show that in this example, for streaming service, using different reference values for different service types actually does impact the ranking. Similarly a comparison of scenarios 1 and 2 shows that the network rankings are impacted if different distribution of attribute weights is used for each service type as opposed to a single distribution of attribute weights for all services.

Finally we evaluate the use of non-monotonic utility for a case where there is one physical access network that supports multiple classes of QoS. An example of that kind of network can be WiFi MultiMedia (WMM)-based access network, i.e., an implementation of IEEE 802.11e standard-based access network with four QoS classes. In such a situation we have to select the most optimal QoS class to deliver the requested service. For cases where real-time information about delay, jitter, and packet loss values can not be obtained from the access network, it may be possible to use static or provisioned values from the service level agreement with the network operator. This would allow decision making with only network utilization requiring a real-time update. Even the utilization rate can in some cases be predicted based on, for example, the past seasonal trends. Table 5 represents such an access system that supports five classes of service or five levels of SLAs. The attribute values are for illustrative purposes for use in the decision process for network selection. For example, the cost attribute in this case is derived based on the treatment the packets from that class of service would get. QoS Class 1, for example, gets least preferential treatment relative to other classes and therefore has the lowest cost. The selection of any particular alternative will map to the same physical network but with a different QoS class. Therefore, while the total access network bandwidth will be the same for all alternatives, the values of other parameters (e.g., delay, jitter, packet loss) are different depending upon the QoS class. The allowed bandwidth has been assumed to be based on operator policy to rate limit its customers using different QoS classes on the access network. The reference networks and attribute weights for different service types are the same as for the previous example. The results that used the three possible implementation of GRA are shown in Fig. 4. As explained earlier, scenario 1 uses a combination of reference values and attribute weights to distinguish among different service types. It therefore provides a more balanced approach with results closest to optimization objectives of the decision maker if the intent is to select the network closest in characteristics to the reference network for that service type. It can be seen that depending upon the implementation of the algorithm, a different QoS class can possibly be selected for service delivery. For example, for a VoIP call, the QoS class of service selected by the preferred GRA

Table 5 Attribute values for alternative networks at the time of network selection

	CB (%)	TB (mbps)	AB (mbps)	U (%)	D (ms)	J (ms)	L (per10^6)
QoS Class#1/SLA#1	10	100	0.1	30	400	100	100
QoS Class#1/SLA#2	20	100	0.5	20	200	50	50
QoS Class#1/SLA#3	40	100	1	20	100	25	50
QoS Class#1/SLA#4	60	100	5	40	50	10	100
QoS Class#1/SLA#1	60	100	1	20	10	5	50

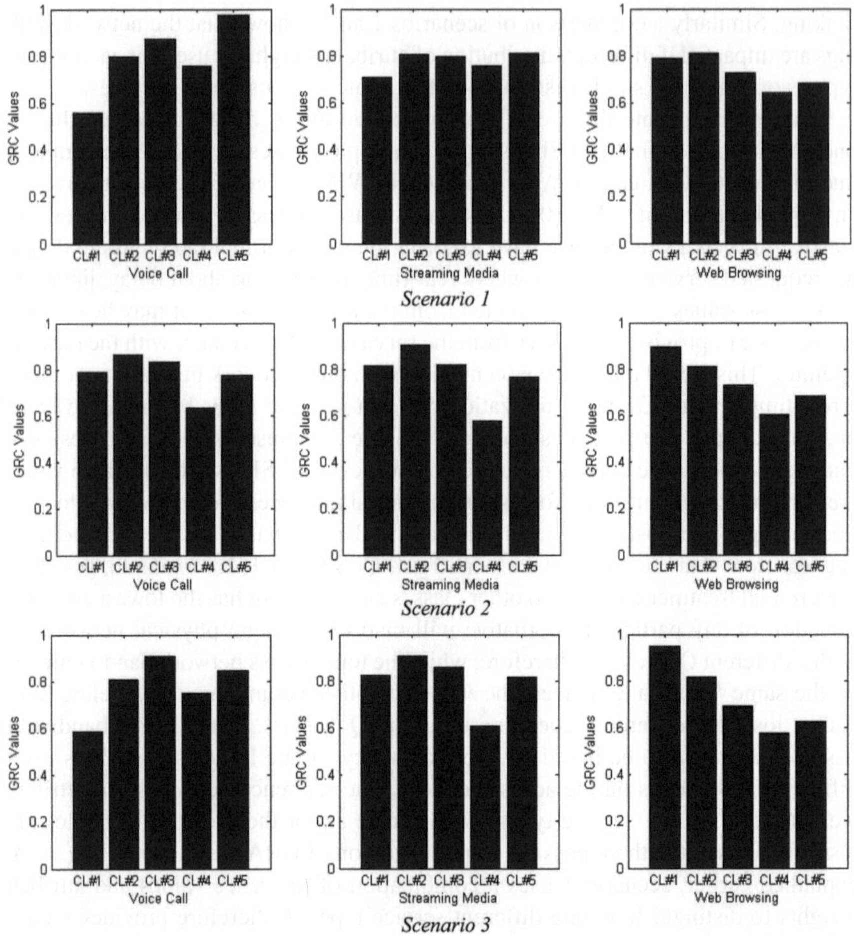

Fig. 4 Results for network selection for three possible configurations of GRA shown in Fig. 2. Configuration 3 is preferred as it provides maximum flexibility to fine tuning the algorithm to optimization objectives

implementation chooses QoS class #3, whereas the other two possible implementations choose QoS class # 2 and QoS class #5, respectively. This again shows that using different reference values for different service types actually does impact the ranking and therefore the service delivery network.

7 Conclusion

Network selection is an important problem to be solved for inter-worked heterogeneous all IP wireless systems. Prior research on application of MADM algorithms to the problem of network selection has not compared them to select the most appropriate algorithm. This chapter has presented the decision maker's optimization objectives and hence the utilities of attributes as a means to evaluate the algorithms and select the most appropriate one. The need to support non-monotonic utility for attributes in order to handle diverse optimization objectives of a decision maker has been shown and MADM algorithms have been evaluated for handling these objectives. We have shown that many of the commonly used MADM algorithms such as SAW, WPM, TOPSIS, and ELECTRE in their standard form are not best-suited because of assumptions about monotonically increasing or decreasing utilities of the attributes. We have also shown that GRA can easily be adapted to use multiple reference networks so that both monotonic and non-monotonic utilities can be taken into consideration, and is therefore better suited for achieving this type of optimization objectives. The evaluation of adapted GRA in this chapter has also demonstrated that the selection of the best-suited delivery network will be impacted by how the algorithm is used to achieve the optimization objectives. A novel two step process that uses multiple reference values has been proposed and explained through an example, which shows how reference attribute values and attribute weights impact the selection process. The adjustment of these parameters for different service types has been discussed. The decision process proposed in this chapter can be used in a heterogeneous wireless network system environment. Future work on the topic could include research into the usefulness of the approach in terms of its effectiveness for network load balancing, business cost savings, and consistency of customer experience.

References

1. Interworking with External Networks Task Group, IEEE 802.11u, http://grouper.ieee.org/groups/802/11/Reports/ tgu_update.htm
2. "Media Independent Handover", IEEE 802.21, http://www.ieee802.org/21/
3. 3GPP Technical Specification Group Services and Systems Aspects; 3GPP System Architecture Evolution: Architecture Enhancements for non-3GPP accesses (2007), 3GPP Technical Report 23.402.
4. Triantaphyllou E (2002), Multi-Criteria Decision Making Methods: A Comparative Study, Kluwer Academic Publishers, Dordrecht.
5. Song Q, Jamalipour A (2004), "Quality of Service Provisioning in Wireless LAN/UMTS Integrated Systems using Analytic Hierarchy Process and Grey Relational Analysis", Proc. IEEE Globecom, TX, USA.
6. Song Q, Jamalipour A (2005), "Network selection in an integrated wireless LAN and UMTS environment using mathematical modeling and computing techniques", IEEE Wireless Commun., 12(3), 42–48.
7. Song Q, Jamalipour A (2005), "A network selection mechanism for next generation networks", Proc. IEEE ICC, Seoul, Korea.
8. Stevens-Navarro E,. Wong VWS (2006), "Comparison between Vertical Handoff Decision Algorithms for Heterogeneous Wireless Networks," Proc. IEEE VTC-Spring, Melbourne, Australia.
9. Hwang CL, Yoon K (1981), Multiple Attribute Decision Making: Mehtods and Applications, Springer Verlag, New York.
10. Benayoun R, Roy B, Sussmann N (1966), "Manual de reference du programme electre, Note de Sythese et Formation," Direction Scietifique SEMA, N. 25.
11. Deng JL (1989), "Introduction to Grey System", J. Grey Syst., 1(1), 1–24.

RFID Indoor Tracking System Based on Inter-Tags Distance Measurements

Abdelmoula Bekkali and Mitsuji Matsumoto

Abstract While in the near future everything will be tagged with Radio Frequency IDentification (RFID) tags, the localization of these tags in their environment is becoming an important feature for many RFID-based ubiquitous computing applications and robotics. Location-aware services in RFID system will allow offering value-added services to the RFID user, and RFID tags can be used for more than just labeling items.

This paper proposes an RSS-based positioning algorithm for objects attached with UHF RFID tags, by means of two mobile RFID antennas and landmarks to overcome the limitations of RFID technology and reduce the localization cost and environment complexity.

The proposed algorithm opens up a possibility for creating novel location-based applications using RFID technology, without specialized hardware or extensive training. It uses an RFID map made from passive or active reference tags (landmarks) to locate analytically any unknown tag detected by the RFID Reader antennas and improve statistically the overall accuracy of locating objects by defining the statistical distribution of the location estimation error for each landmark.

This algorithm is independent from the readers coordinates, and hence it can be more practical due to its mobility and its low cost to achieve a high deployment. To minimize the effect of the RSS and the process measurement noises on the position estimation, an adaptive Kalman filter and probabilistic map matching are applied.

Results obtained after conducting extensive simulations demonstrate the validity and suitability of the proposed positioning algorithm to provide high-performance level in terms of accuracy and scalability.

A. Bekkali (✉)
Graduate School of Global Information and Telecommunication Studies, Waseda University, Japan
e-mail: bekkali@fuji.waseda.jp

S. Powell, J.P. Shim (eds.), *Wireless Technology*, Lecture Notes in Electrical Engineering 44, DOI 10.1007/978-0-387-71787-6_3,

1 Introduction

In recent years, Radio Frequency IDentification (RFID) technology has moved from obscurity into mainstream applications that help speed the handling of manufactured goods and materials. RFID is a means of storing and retrieving data through electromagnetic transmission to an RF compatible integrated circuit. It fundamentally consists of two elements: the transponder which is located in the object to be identified and the reader [1]. RFID system has been widely adopted as an attractive technology for many significant applications such as asset tracking, industrial automation, and homecare, and health-care systems.

Location finding systems for indoor areas are an emerging technology that has become very important in recent years. The location information is one of the most important and frequently used contexts in ubiquitous computing [2]. Such system can use the changes of location to adapt its behavior, such as computation and communication, without user intervention.

RFID tags can be used for more than just labeling items. However, the main limitation involved is that the spatial location information of an object cannot be acquired through the current RFID technology. The ability to determine the spatial location of units belonging to a RFID technology is a starting point toward the development of sophisticated applications, such as people tracking for civil protection, patients monitoring in hospitals, and quick rescuing of victims [3].

Localization is usually associated with global positioning system (GPS), which can achieve localization accuracy up to about 3 m. Unfortunately, we cannot use this classic location technology indoors because the radio signals sent by GPS satellites are too weak to penetrate walls [4]. In order to overcome the disadvantage of GPS and locate objects accurately in complicated indoor environment, a lot of research has been conducted and different solutions have been proposed over the years (Ni et al., 2003, Personal Communication) [2, 5]. These solutions differentiate themselves on the basis of the method used, processing time, accuracy, hardware, and deployment cost. RADAR [5] is an RF in-building tracking system based on the IEEE 802.11WLAN. It was the first system to propose the use of a RF map of the area. Received Signal Strength (RSS) for each WLAN base station is stored as a fingerprint in a database for each point in a dense grid covering the floor. When querying the database, a nearest neighbor match in the fingerprint space provides candidates for mobile's position. LANDMARC (Ni et al., 2003, Personal Communication) is an RFID-based positioning scheme that is in a way similar to RADAR scheme, except that the RF map is built by previously placed active tags.

The development of an efficient and accurate location sensing systems for indoor environments, based on the current RFID system, is a challenging task. Indeed, RFID is a technology constrained by many differing environmental factors, of which need to be considered, before a successful localization system can be commissioned. The limitations usually stem from the harsh nature of the RF signal (absorption, scattering, multipath, etc.) and other factors related to the technology itself such as the antenna orientation and polarization matching for both the reader and the tag and the effect of RFID tag placement. In most cases, the orientation of the reader and tag antennas cannot be precisely matched, causing loss in transmitted power. This can lead to unpredictable read range even in environments that are free of material and radio interference. However, using circular polarization in the reader antenna can minimize this problem of orientation. Antenna polarization can cause power loss in the link budget, and its effects must be understood in a successful RFID environment [1, 8]. Furthermore, the lossy dielectric or metallic surfaces, on which the RF tags are placed and/or nearby objects and materials, can affect the tag's performance. It may improve the performance by directing the reflected signal toward the system antenna or it may decrease performance by reflecting the signal away from the system antenna or by absorbing a portion of the signal [6–8]. These effects make it infeasible to construct a simple and accurate model of the RFID signal's propagation indoors. The tracking system has to overcome the high uncertainty due to the behavior of the indoor wireless channels but at the same time it should keep the cost and the complexity of large-scale deployment as small as possible.

In addition, the RADAR [5] requires a large number of RSS data stored in a database (fingerprint) defined in online mode to achieve the localization, which is difficult to design and does not take in consideration the environment or the access point location changes. Further, to estimate the tag location, the LANDMARC (Ni et al., 2003, Personal Communication) requires at least three readers installed in order to have a good geometry, which make the system costly. Moreover, due to the limitation of RFID antenna, the detection of the tag by three readers at the same time is difficult in most cases.

The proposed approach aims to provide a low cost and low complexity tracking system based on RFID technology in smart environment, where everything will be tagged with RFID tag. By considering that the RSS value is embedded in the RFID Readers, the location of RFID tag can be estimated by mean of mobile reader using only RSS measurement without any additional hardware. The RSSI-based approach is appealing for its low cost, even if the position estimate of the target is not particularly accurate. In the other hand, the existing positioning systems (Ni et al., 2003, Personal Communication) [2, 5] require pre-organization of location of sensors or Access points, thus the setup and the training cost might limit their applicability in a large-scale deployment. Moreover, these approaches had the problem of the number and placement of stations which affect the location estimation accuracy.

This chapter consists on designing an indoor tracking system for mobile computing applications that can provide position estimates, without the need of specialized hardware and extensive training by the following:

- Proposing an analytical algorithm that provides location estimation of the target using mobile RFID readers and multiple RFID location reference points (landmarks) which furnish data for localization without also the use of inertial sensors. The positioning algorithm is independent from the readers coordinate to ensure the mobility of the readers.
- Designing a probabilistic RFID map made from statistical distribution of the location estimation error for each landmark. The main purpose of the probabilistic RFID map is to model how the location measurement error is distributed in

different geographical areas based on a sample of location measurements collected online at several known locations inside the room and help to improve the accuracy of the target location.

This chapter is organized as follows. In Section 2, we describe theoretical background of indoor propagation model for far-field backscattered RFID system. Section 3 presents the details of our indoor tracking algorithm. In Section 4, we evaluate the performance of the proposed algorithm. Finally, Section 5 concludes the chapter and provides directions for future research.

2 Indoor Propagation Model for Far-Field Passive RFID System

The changes in signal due to propagation indoors are difficult to predict because of dense environment and propagation effects such as reflection, diffraction, and scattering. However, the received signal is usually modeled by the combined effects of large-scale fading and small-scale fading. Theoretical and empirical models indicate that the average received signal power decreases log-normally with distance for indoor wireless channels.

There are other channel impairments that can degrade link performance. These impairments include delay spread due to multipath and fast fading (distortion of the signal spectrum) [9]. These effects must be considered by the equipment designer, but are not generally considered as part of communication link design.

For localization applications, the modeling of the propagation channel is for the purpose of predicting the RSS at the end of the link and thus the distance between the transmitter and the receiver.

In this section, we briefly discuss the large-scale fading component for RF-modulated backscatter system which describes the signal attenuation as the signal travels over a distance and is absorbed by material such as walls and floors along the way to the receiver. Here, we will not consider the tag antenna gain loss due to material attachment, which can be part of the path loss model as described in [6].

2.1 Backscattered Free-Space Link Budget

The signal transmitted on the downlink (reader to the tag) contains both continuous wave (CW) and modulated commands as shown in Fig. 1. On the uplink (tag to the reader), the data are sent back during one of CW periods when the tag impedance modulates the backscattered signal. More details on RFID system protocol and operation can be found in [10, 11, 12].

Hence, in free space, the link budget of the modulated backscatter communication, which is known also as *modulated radar cross-section*, can be expressed as [13, 14, 9].

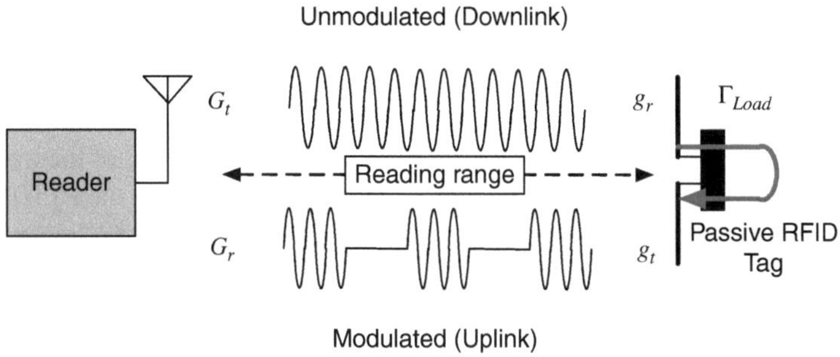

Fig. 1 Far-field-modulated backscatter communication links

$$\frac{P_r}{P_t} = \underbrace{\left[G_t g_r \left(\frac{\lambda}{4\pi d_0}\right)^2\right]}_{Downlink} \Gamma \underbrace{\left[G_r g_t \left(\frac{\lambda}{4\pi d_0}\right)^2\right]}_{Uplink} \tag{1}$$

where P_t and P_r are the powers transmitted and received by the reader transmit and receive antennas, G_t, g_t and G_r, g_r are the gains of the reader and tag transmit and receive antennas, respectively, and Γ is a reflection coefficient of the tag.

Further rewriting yields to

$$\frac{P_r}{P_t} = G_t G_r \left(g_t \Gamma g_r\right) \left(\frac{\lambda}{4\pi d_0}\right)^4 \tag{2}$$

where the quantity $(g_t \Gamma g_r)$ is defined as function of radar cross-section δ [14, 15] and the wavelength λ

$$(g_t \Gamma g_r) = 4\pi \delta / \lambda^2 \tag{3}$$

2.2 Indoor Large-Scale Propagation Model

In order to estimate the distance between a transmitter and a receiver, the log-distance path loss model that has been used extensively in the literature [9] is used:

$$PL(d) = PL(d_0) + 10n \log\left(\frac{d}{d_0}\right) + X_\sigma \quad [dB] \tag{4}$$

where d_0 is an arbitrary reference distance (usually 1 m), $PL(d_0)$ is the free-space path loss for distance d_0, n is the path loss exponent and X_σ is zero mean Gaussian random variable with variance σ_{dB}^2 in dB. The variable X_σ called the shadow fading and used to model the random nature of indoor radio signal propagation due to the effect of various environmental factors such as multipath, obstruction, tag orientation, and moving objects between the transmitter and the receiver.

In modulated backscatter scheme, the path loss measurements confirm that the path loss exponent of the two-way link is approximately twice that of the traditional one-way link in the same environment [13]. Thus, the path loss in the RFID passive communication may be expressed as

$$PL(d) = PL_0 + 10N\,\log(d) + X_\sigma \quad [dB] \tag{5}$$

where PL_0, defined by (2), is the path loss in the free space at distance 2 m, $N = 2n$ is the path loss exponent.

3 The Proposed Algorithm

In this section, we introduce an indoor location sensing system for RFID-tagged objects in dynamically changing environment. The research challenge corresponds to achieving an accurate indoor tracking system using mobile RFID reader with unknown location and landmarks. Hence, the reader can move until it detects the target to start the localization process. The proposed approach is based on RSS measurement to measure the reader-tags distance and target-landmarks distance to estimate the target location. In general, RSS is an ideal modality for range estimation in wireless networks because RSS information can be obtained at no additional cost with each radio message sent and received. To mitigate the limitations of RF and RFID technology described earlier and the process measurement noise on the location estimation, we design a probabilistic RFID map made from statistical distribution of the location estimation error for each landmark. This statistical map serves to model the location estimation variation inside the room and help to improve the accuracy of the target location.

The proposed algorithm consists of three main parts: (1) the localization method which is based on inter-tags distance measurements, (2) the probabilistic error map generation, and (3) the Adaptive Kalman filtering. Figure 2 describes the general block diagram of the proposed positioning algorithm.

3.1 Inter-tags Distance Measurement Algorithm

This algorithm uses a map generated from passive or active references tags with known location (landmarks) to locate any unknown target detected by the RFID Readers (Fig. 3). It measures the distances between the RFID readers and the common detected tags using the large-scale path loss propagation model described above and calculates the distance between the target and all the detected landmarks. With the multilateration technique the system is able to estimate the position of the unknown tag.

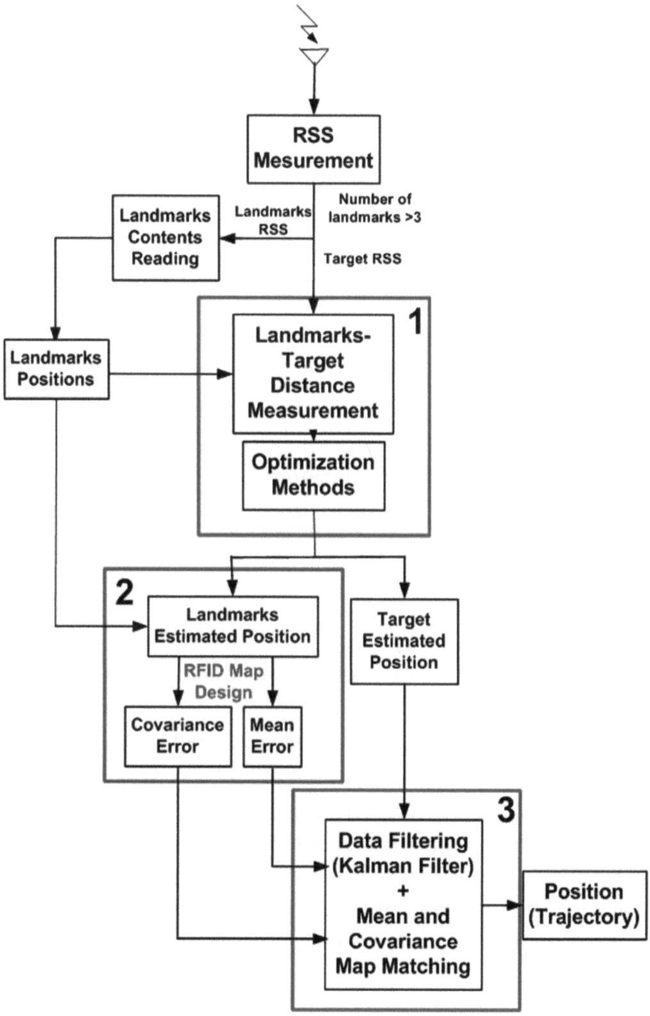

Fig. 2 Block diagram of the RFID indoor positioning algorithm

The concept behind this algorithm is to estimate the target's location with reference to the N RFID landmarks, not to the two RFID readers coordinates. Therefore, the location estimation is done between elements in the same environment with the same error correlation. As more landmarks are available we can obtain more distance estimates that can be used to form an overdetermined system (multilateration) and can provide better accuracy than using only three readers (trilateration).

Let us assume that

$\left|d_i^{(k)}\right| i = 1,..N;\ k = 1,2$ is the estimated distance between the ith landmark and the readers R1 and R2.

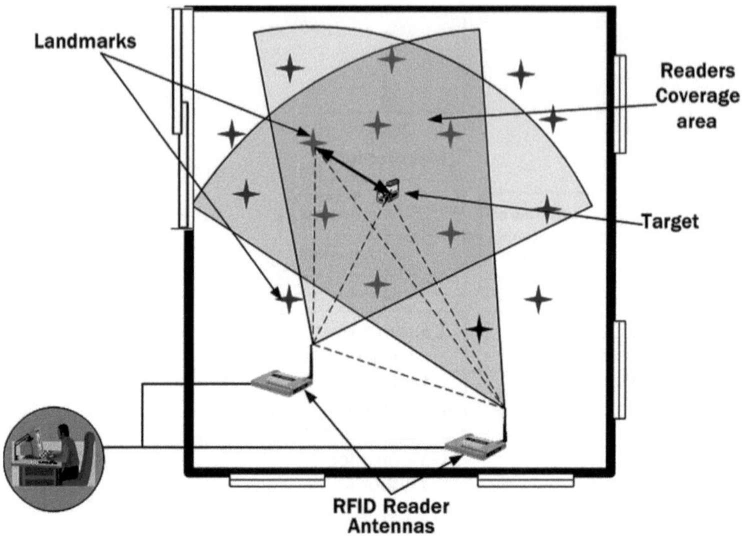

Fig. 3 RFID indoor positioning scenario

$\left|d_T^{(k)}\right| k = 1{,}2$ is the estimated distance between the unknown tag and the two readers.

$|d_{12}|$ is the distance between the readers R1 and R2.

Since the proposed approach is based on received signal strength (RSS) measurement, the distances between the readers and the RFID tags can be estimated using the Eq. (1).

The distance between the tag and the *i*th landmark is illustrated in Fig. 4. Note that the Fig. 4 can be seen as a reference $\Re(R_1, R_1R_2, R_1P_i)$ in which we define the position of the tag T.

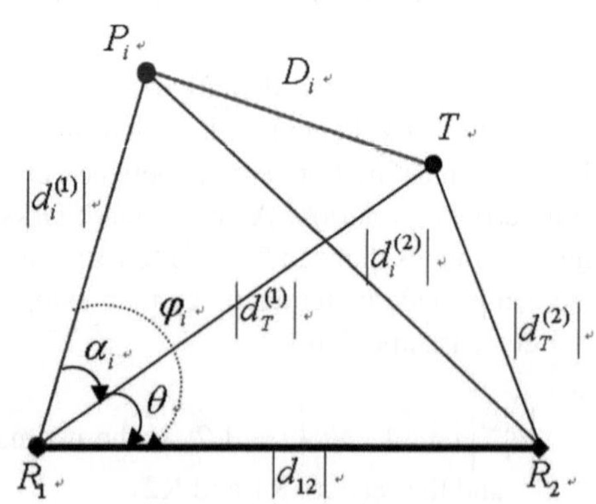

Fig. 4 Tag-landmark distance measurement

Applying the law of cosine, the distance D_i between the landmark P_i and the tag T can be obtained from this set of equations:

$$D_i = \sqrt{\left|d_i^{(1)}\right|^2 + \left|d_T^{(1)}\right|^2 - 2\left|d_i^{(1)}\right|\left|d_T^{(1)}\right|\cos(\varphi_i - \theta)} \tag{6}$$

where

$$\varphi_i = \cos^{-1}\left(\left(|d_{12}|^2 + \left|d_i^{(1)}\right|^2 - \left|d_i^{(2)}\right|^2\right) \Big/ 2\,|d_{12}|\left|d_i^{(1)}\right|\right) \tag{7}$$

$$\theta = \cos^{-1}\left(\left(|d_{12}|^2 + \left|d_T^{(1)}\right|^2 - \left|d_T^{(2)}\right|^2\right) \Big/ 2\,|d_{12}|\left|d_T^{(1)}\right|\right) \tag{8}$$

Using these relations for each landmark we can obtain a set of distances $\{D_i, i = 1, \ldots N\}$ between the each landmark and the target to be located. Since we know the correct position of the landmarks P_i, the distance D_i can be expressed as the distance between the correct position P_i and the target X:

$$\|P_i - X\|_{P_i} = D_i \quad \forall i = 1, \ldots N \tag{9}$$

If we denote the coordinates of the ith landmark as (x_i, y_i) then the location of the target (x, y) can be obtained by solving the system of non-linear equations of the form:

$$(x_i - x)^2 + (y_i - y)^2 = D_i^2 \quad \forall i = 1, \ldots N \tag{10}$$

To solve the system of non-linear equations we treat it as an optimization problem and apply the minimum mean squared error algorithm:

$$(\hat{x}, \hat{y}) = \min_{(x,y)} \sum_{i=1}^{N} (f_i(x, y))^2 \tag{11}$$

where $f_i(x, y)$ is the estimation error and is given by

$$f_i(x, y) = \left|\sqrt{(x_i - x)^2 + (y_i - y)^2} - D_i\right| \tag{12}$$

This optimization problem needs to solve a set of non-linear equations and therefore requires a large number of repetitive numerical computations especially when we have to repeat this algorithm with several RSS sample to get better solution. However, such calculations are not suitable for real-time application. To obtain a suitable algorithm for accurate estimation in real-time, Eq. (11) may be linearized and the solution can be calculated using least squares estimation (LSE).

3.2 Probabilistic RFID Map Generation

In the previous section, we described an analytical algorithm to estimate the location of the target using landmarks. This location estimation method suffers from severe errors, due to the RSS measurement data from mobile RFID antennas and others described earlier. To enhance the location estimation of the target, we model the effect of the measurement error on the location estimation for each landmark, which are distributed in order to cover the whole room. Hence, we can get an RFID Map of the location estimation error (LEE) for the antenna's detection area or the room.

The proposed model is based on the knowledge of statistical distribution of the LEE for each landmark to imitate the location measurement error inside the reader's coverage area. Using statistical parameters of the LEE model, the target location estimation can be enhanced by utilizing a probabilistic matching technique. Usually, the statistical modeling of the environment should be the good choice where deterministic methods fail to imitate the error environment. To design the LEE model inside the reader's coverage area, we apply the localization algorithm described in Section 3.1 to each landmark. As we can read the true location of the landmarks by the readers, we can have the probability distribution function (*pdf*) of the localization error at each landmark (Fig. 5) by calculating the expectance (mean) and the variance of the measured data using maximum likelihood estimator (MLE) [19]. The location estimation error is Gaussian comes from the empirical data of the RSS which suggest that RSS distribution for a fixed location inside the room is Gaussian [9]. Having estimated the LEE parameters (mean and variance), the location estimation of the target can be filtered using adaptive Kalman filtering.

Therefore, the RFID map can be defined as follows:

$$RFID\_Map \hat{=} \{(l_i, pdf(i); \quad i = 1,..N\} \tag{13}$$

where

N is the number of landmarks.
l_i is the true location of the ith landmark.

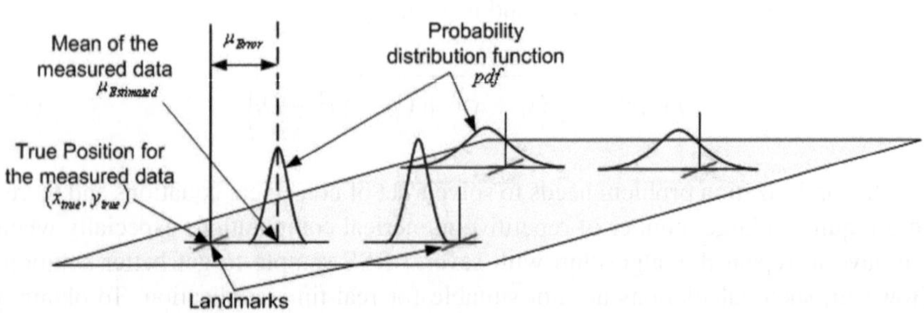

Fig. 5 Probabilistic RFID map generation concept

$pdf(i) \sim N(\mu_{error}(i),\sigma^2_{error}(i))$ is the *pdf* of the measurement error which is supposed to be Gaussian.

This method in a way is similar to that used by WLAN known as fingerprinting [5], but instead of storing the RSS at the corresponding locations in a database defined in online mode, we will store directly the location in the landmarks and create online RFID map which is better in term of mobility and reliability.

3.3 Adaptive Kalman Filtering

Despite the existence of more sophisticated filters, Kalman filtering (KF) has been used successfully in different prediction applications or state determination of a system. It is the one that minimizes the variance of the estimation error. For a detailed description of the Kalman filter see [16]. The Kalman filter addresses the general problem of trying to estimate the state $x \in \Re^n$ of a discrete-time controlled process that is governed by the linear stochastic difference equation:

$$x_{k+1} = Ax_k + Bu_k + w_k \tag{14}$$

With a measurement (observation) $y \in \Re^m$ that is

$$y_k = h(x_k) + e_k \tag{15}$$

The random variables $w_k \sim N(0, Q_k)$ and $e_k \sim N(0, R_k)$ represent the process and measurement noise (respectively). A and h have to be chosen in order that they model as properly as possible the motion behavior of the target.

The implementation of KF requires the a priori statistical knowledge of the process noise Q and the measurement noise R. Poor knowledge of the noise statistics may seriously degrade the KF performance and even provoke the filter divergence [17]. To fulfill the requirement, an adaptive Kalman filter can be utilized by estimating the noise covariance matrices. Here, we will consider only the measurement noise covariance R to adapt Kalman filter.

3.3.1 Motion Model

In order to use Kalman filter to remove noise from a signal, the process that we measured is described by the linear system Eq. (14). In the following, ^{T} denote the transpose of a vector or a matrix. Define a (4-D) stochastic process $x(t) = (x_1(t),x_2(t),v_1(t),v_2(t))^T \quad t \in \Re$. $x_1(t),x_2(t)$ denote the x and y coordinates of the mobile target and $v_1(t),v_2(t)$ the x and y coordinates of the velocity vector at time t. Observation are taken at discrete time point $t_k = t_0 + \Delta t.k, \quad k = 1..N$.

In our model, we assume that the mobile's motion can be modeled by

$$p_{k+1} = p_k + \Delta t v_k + \frac{\Delta t^2}{2} a_k \tag{16}$$

$$v_{k+1} = v_k + \Delta t a_k \tag{17}$$

where $p = (x_1(t),x_2(t))^T, v = (v_1(t),v_2(t))^T$ and a is the motion acceleration which define the velocity variation. If we assume the motion is uniform. Thus, the matrix A and state covariance error Q of the process equation Eq. (14) can be defined as follows:

$$A = \begin{pmatrix} 1 & 0 & \Delta t & 0 \\ 0 & 1 & 0 & \Delta t \\ 0 & 0 & 1 & 0 \\ 0 & 0 & 0 & 1 \end{pmatrix} \quad B = \begin{pmatrix} \Delta t^2/2 \\ \Delta t^2/2 \\ \Delta t \\ \Delta t \end{pmatrix} \quad Q = \sigma_a^2 \begin{pmatrix} \frac{\Delta t^4}{4} & 0 & \frac{\Delta t^3}{2} & 0 \\ 0 & \frac{\Delta t^4}{4} & 0 & \frac{\Delta t^3}{2} \\ \frac{\Delta t^3}{2} & 0 & \Delta t^2 & 0 \\ 0 & \frac{\Delta t^3}{2} & 0 & \Delta t^2 \end{pmatrix} \tag{18}$$

where σ_a is the acceleration error.

In the case where the target is fixed, we assume that the process noise, which is the noise in the location estimation itself, is negligible when compared to the measurement noise. Also the state of this system is not modified with any control input ($B = 0$). With these assumptions, the Eq. 14 can be simplified as follows:

$$x_{k+1} = x_k \tag{19}$$

3.3.2 The Observation Model

The observation model describes how measurements are related to the states. The KF needs a model of the measurements in order\Aq[531]{Please conform whether this eqn is numbered or unnumbered} to correct the state prediction when an observation is available.

The observations $\{y_k, k = 1,..M\}$ consist of the measured location of the fixed target for each received RSS data (M samples). In general, the random variable $\{y_k, k = 1,..\}$ is Gaussian with arithmetic mean $\bar{y}$ and variance σ_y^2. Moreover, if we know the correct location of the measured data y_{true}, the resulted estimated data are shifted with an error μ_{error}. Hence, $\bar{y}$ can be defined as follows:

$$\bar{y} = y_{true} + \mu_{error}.$$

Therefore, for each process step, the measurement equation Eq. (15) can be modeled with a linear form as follows:

$$y_k = h(x_k) + e_k = Hx_k + \mu_{error}^k + e_k \tag{20}$$

where

$$H = \begin{pmatrix} 1\,0\,0\,0 \\ 0\,1\,0\,0 \end{pmatrix}; \quad E(e_k e_i^T) = \begin{cases} R_k, i = k \\ 0, i \neq k \end{cases} \tag{21}$$

$\mu_{error}^k = x_k^{true} - E(y_k)$ defines the mean of the location measurement error, while x_k^{true} is the true position of the target.

To estimate (μ_{error}^k, R_k) for each process step, we use the RFID map defined in the Section 3.2 by finding the nearest landmark for each new measurement y_k. The basic idea on the mean and covariance matching approach is to apply the measurement noise parameters measured from the nearest landmark to the measurement noise parameters of the target. Thus, better resolution of landmarks leads to better estimation of the measurement error of the target.

The Kalman filter prediction and correction steps with the mean and covariance matching algorithm are illustrated in Fig. 6

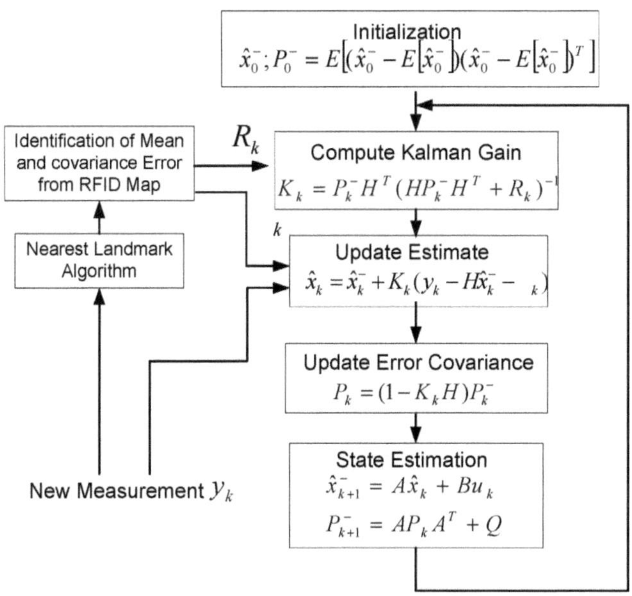

Fig. 6 Flowchart for RFID map-aided adaptive Kalman filter

4 Simulations and Evaluation

The major performance metrics of interest for indoor positioning systems are the accuracy and the precision in estimating a position. In this section, we evaluate the performance of our proposed positioning system using passive devices and investigate how the path loss, RSS characteristics and the number of landmarks influence the precision. We will see how the accuracy can be improved by using probabilistic map and Adaptive Kalman Filtering. As the algorithm is based on mobile RFID positioning, its performance is compared with fixed RFID readers with known location.

4.1 System Model Setup

The simulations have been performed using Matlab [18]. Recognizing the fact that the accuracy of the indoor localization, based on landmarks randomly distributed, will depend not only on the positioning algorithm but also on the Geometry Dilution of Precision (GDOP) of the landmarks [4], it will be considered that the landmarks are placed in grid form (Fig. 7). For simulation purposes, we assume that we have two RFID readers and placed in the corner of the room at positions (0,0) and (5,0) and the common coverage area (gray color) in which the readers can detect the same tags (target and landmarks) is defined by $S = \{(x,y);\quad 1 \le x \le 4,\ 1 \le y \le 4\}$.

The in-building path loss model chosen for the purpose of simulation is given in (1) and is explained by Rappaport [9]. To this aim, we assumed that the operating frequency is 915 MHz, the power of interrogation signal is 30 dBm, TX/RX antennas gain of the Reader and the tag is 0 dBi (Omni directional antennas), and radar cross-section loss is –20 dBm. Finally, for the propagation model parameters,

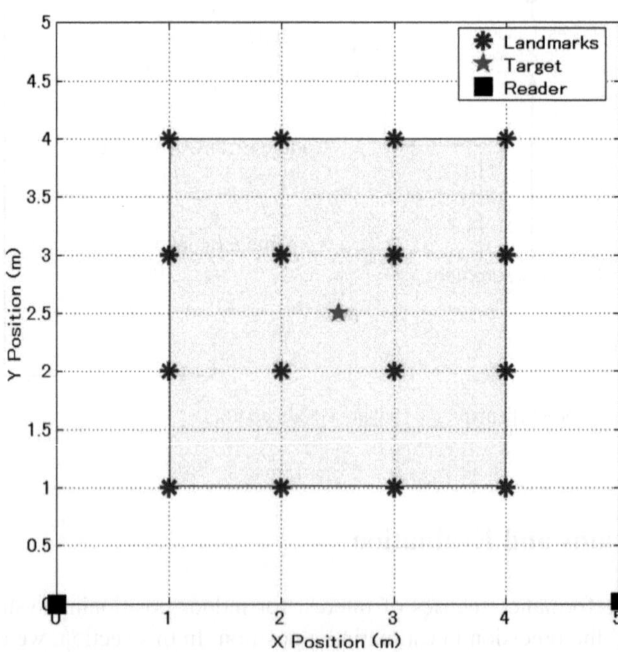

Fig. 7 Landmarks grid distribution for indoor positioning

the backscattered power measured by the reader is affected by zero mean Gaussian random noise with standard deviation $\sigma = 5.2\,dB$ and the path loss exponent n is assumed to be 1.8 as reported for grocery store by Rappaport [9]. Five hundred RSS sample values are generated for each data point.

4.2 Probabilistic RFID Map

To know the effect of the RSS measurement error on the position precision, we analyze how the position error is distributed inside the room, before filtering and correction are applied. Figure 8a shows an example of the resulted probabilistic RFID map using four landmarks $\{(1,1), (1,4), (4,1), (4,4)\}$ and the position estimation distribution of target located at (2.5,2.5) with an RSS error $\sigma_{RSSI} = 2\,dB$. For each landmark a *pdf* of the location estimation is plotted. As can be seen, the mean error is deviated from the true position. The results suggest that the estimation error in term of variance and expectance increases as the distance from the readers increases. Hence, the measurement error is not uniform inside the reader's coverage area. An optimal resolution or distribution of landmarks and/or the short distance between the readers and the target may increase the location accuracy. Figure 8b shows that the probability distributions for the X-axis and Y-axis of the position are Gaussian.

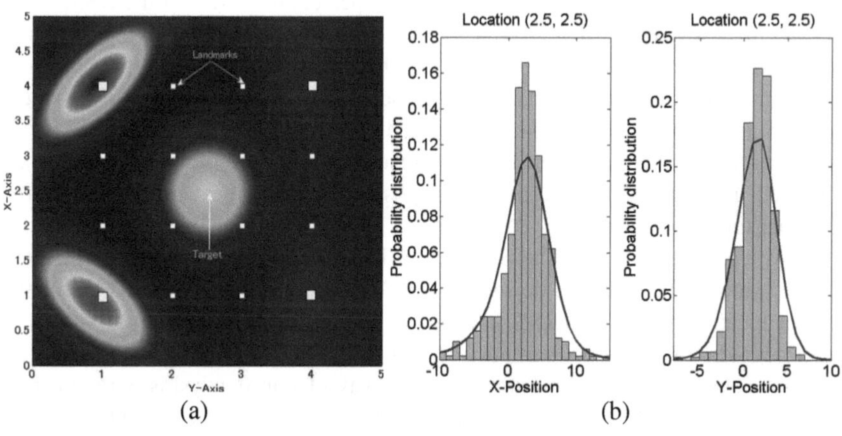

Fig. 8 (**a**) Probabilistic RFID map and (**b**) histogram of the location estimation

4.3 Localization Performance Evaluation

From Eq. (6), the localization algorithm depends on the separation distance between the reader antennas. To know the impact of the separation distance of the reader's antennas and the role of Kalman filter and RFID map matching, we change the

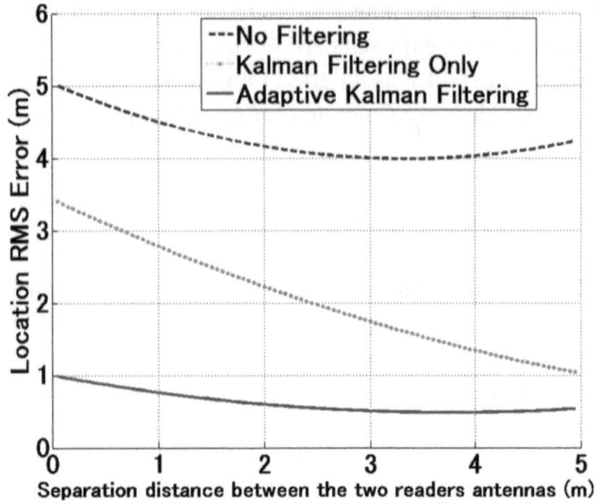

Fig. 9 Effect of separation distance between the two readers antennas

separation distance while keeping the *X*-axis of the target in the center of the two readers. Figure 9 shows the average in the least squares (LS) sense of root mean square (RMS) of the location measurement error of the target at the position (2.5,2.5) with grid spacing 1 m (16 tags). As expected, the effect of the RSSI noise measurement with $\sigma_{RSSI} = 5.2\,dB$ has been reduced dramatically with the combination of Kalman filtering and RFID map matching. On the other hand, the proposed algorithm is less sensitive from the readers separation distance. Hence, the readers can be moved, until the target can be detected by the two antennas to start the localization process, which not the case if the readers are fixed. Note that the proposed approach implements the filtering at the position measurement noise, not for the RSS measurement noise.

Now, we investigate how the accuracy of location estimation would be affected by the value of the standard deviation of RSS. Figure 10a plots the RMS for the correct location versus the RSS standard deviation. Clearly, the larger values of σ_{RSSI} degrade the accuracy dramatically. However, this value is difficult to control because it depends on the environment. One way of improving this is to consider many RSS samples. Results for the localization RMS error as function of the number of samples and σ_{RSSI} are shown in Fig. 10b. The localization accuracy improves as the number of samples increases and the need for more samples increases when the RSSI standard deviation σ_{RSSI} is bigger. This can explained as follows: When the number of sample is sufficiently large, the sampling distribution of the target location is well approximated by a normal distribution, even when the target location distribution is not itself normal (*Central Limit Theorem*) [19]. Thus, larger number of samples makes the mean and variance used in Kalman Filter and RFID map matching more accurate.

We next consider the impact of the grid spacing and the number of the landmarks on the localization accuracy. Figure 11 shows that large landmarks grid spacing

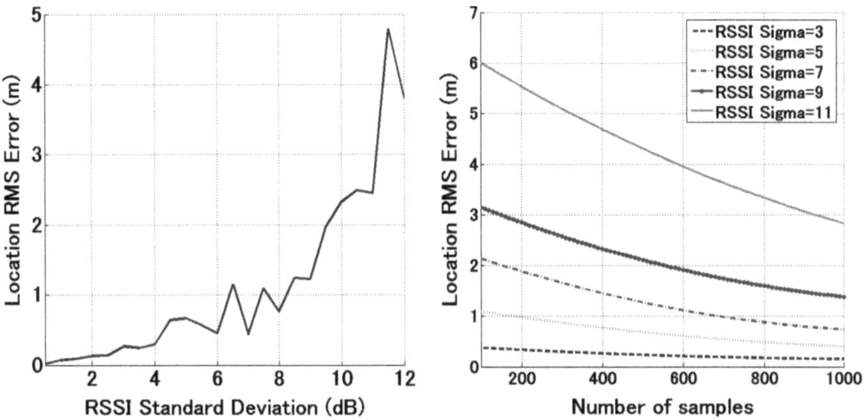

Fig. 10 Effect of standard deviation of RSS Gaussian noise and RSS samples

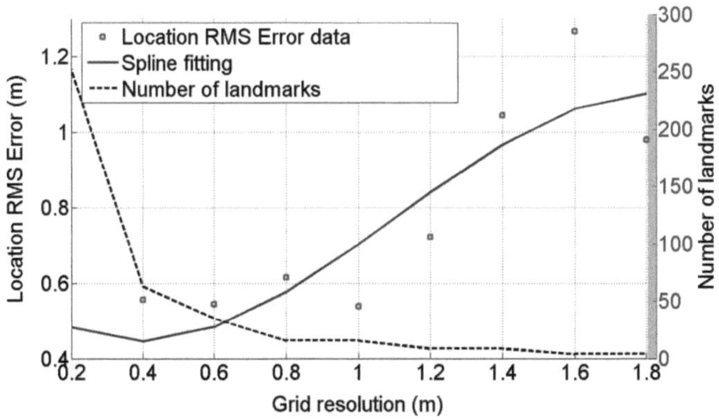

Fig. 11 Effect of number of RSS samples and number of landmarks

leads to poor accuracy. However, the RMS error does not increase significantly as the grid spacing (landmarks number) increases; this is due to the redundancy of the location measurement which is based on multilateration with the landmarks (co-linearity of the landmarks with the target), and also the RSS measurement error is assumed to be the same in the whole room ($\sigma_{RSSI} = 5.2\,dB$).

4.4 Tracking Performance

To analyze the performance of the tracking algorithm inside the readers' common coverage area S (Fig. 7), we simulate of the target's motion as shown in Fig. 12 **.** The details of the target motion parameters are given in Eqs (16) and (17). We considered for simplicity that the commanded acceleration is null and the acceleration noise is $0.1\,m/s^2$, while the target position is measured each second $\Delta t = 1\,ms$.

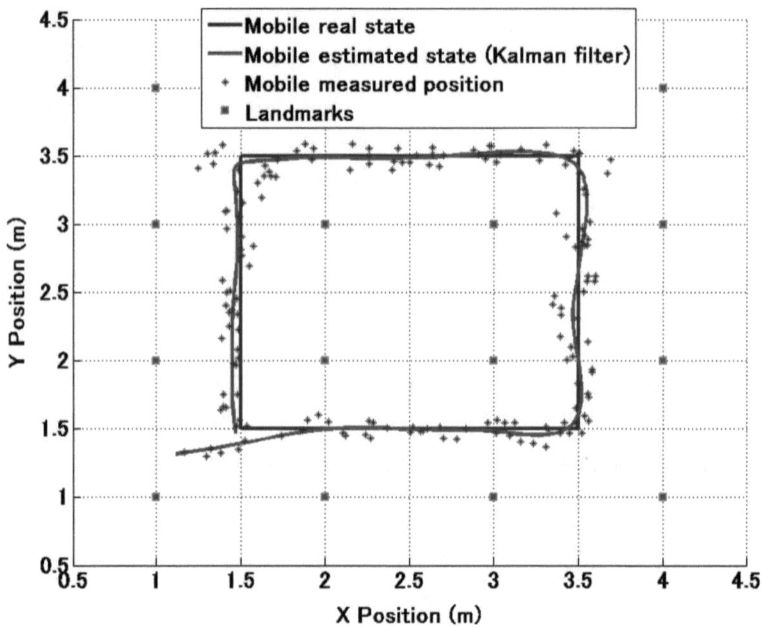

Fig. 12 Performance of indoor tracking algorithm

The Fig. 12 plots the trajectory of the target with reference to the real trajectory. As expected, the effect of the RSS noise measurement with $\sigma_{RSS} = 5.2\,dB$ has been reduced dramatically using Kalman filtering and RFID map matching. Figure 13 illustrates the role of RFID map matching on the tracking accuracy. Therefore, Kalman filter is used to improve the precision by removing the measurement noise variance and map matching to improve the accuracy by adjusting the measurement bias.

Because our proposed algorithm is based on two readers with unknown location and landmarks, we compare its performance with the positioning method based on two readers with known location, which is an analytical intersection of two circles. We apply Kalman filtering to its resulting data and we assume that only one possible solution will be inside the room. Figure 14 shows the average in the least squares senses the RMS for the location measurement error of the target at the locations (2.5, 2.5), while Fig. 15 plots the trajectory of the target using the same tracking assumptions described earlier. The results suggest that our proposal provides better accuracy comparing with the method that uses two readers with known location, when the distance between the readers and the target is bigger and when the separation distance between the two readers is smaller. This can be intuitively explained as follows. If the readers are close together the overlapping region or the uncertainty area becomes large. Generally the localization with only two readers depends on the precision of the distance measurements and the geometric configuration of the readers' position. Furthermore, our proposed algorithm is based on landmarks for

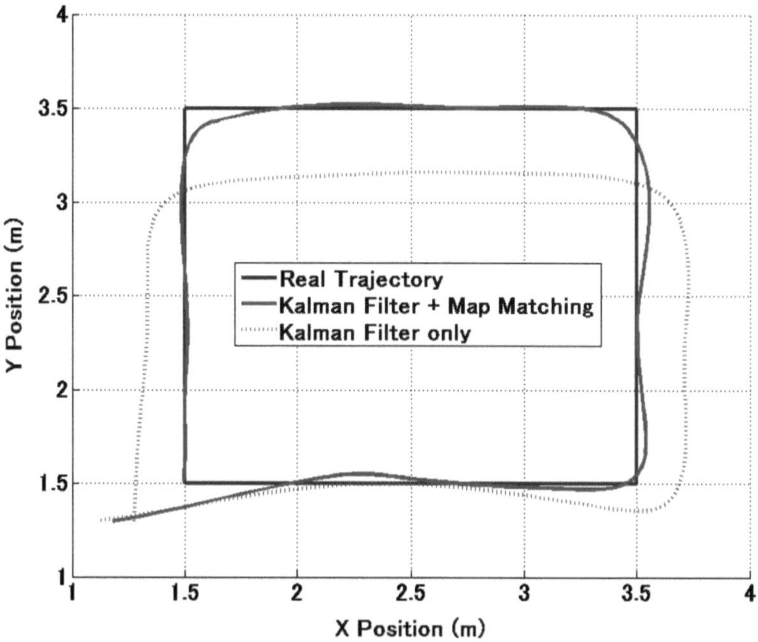

Fig. 13 Impact of RFID map matching on the location accuracy

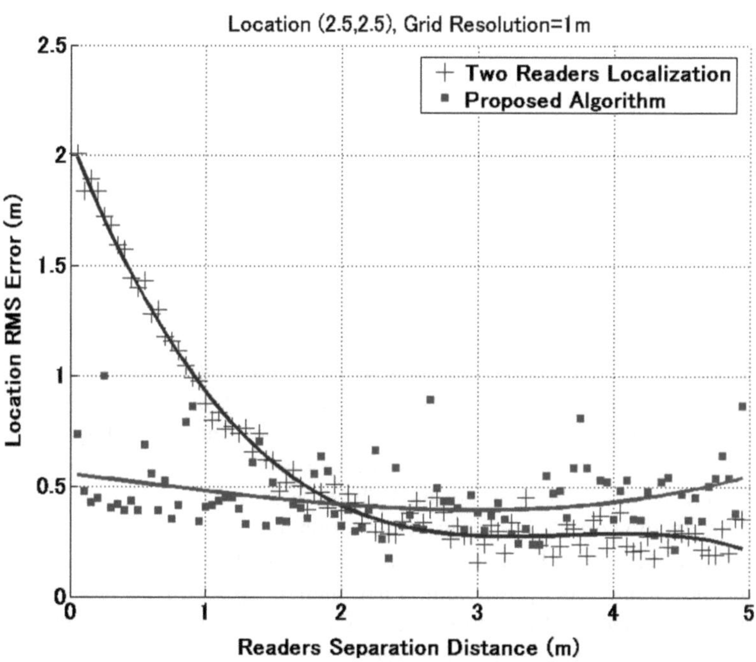

Fig. 14 Average RMS comparison between the two reader localization and the proposed algorithm

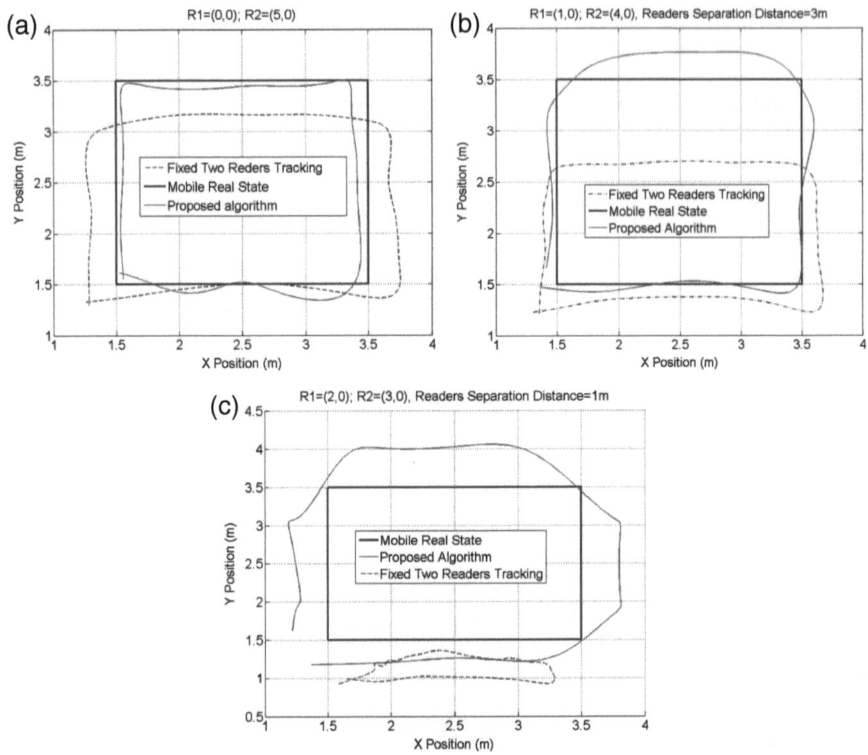

Fig. 15 Performance tracking compared with two readers with known location

localization and correction. When the localization error is high the target's location converges to the position of the nearest landmark (Fig. 15c).

5 Conclusions

Toward the goal of adopting RFID system as a good solution for indoor localization, different aspects of the problem of location determination in RFID system have been investigated. Also a low complexity and low-cost indoor tracking algorithm, by means of two mobile RFID readers and landmarks, has been introduced. The main features of this algorithm are the utilization of adaptive Kalman filter and a map of localization errors to improve the positioning.

The proposed approach has two major advantages that can widely be used for tracking and mobile robot positioning. First, there is no need for a large number of expensive RFID readers. In addition, the system estimates the target location, using multilateration with reference to the landmarks (RFID map). Hence, is independent from the readers' coordinates and their separation distance. The readers can move until the target is detected and start the localization process and also while moving the readers toward the target, stronger RSS can be measured. Second, we defined an online database of the probability distribution function (*pdf*) of the location estimation error for each landmark to correct the location estimation of the target, instead of using RF fingerprinting which should be defined in online mode. Therefore, it will be less complex and immune to the time variations of the environment.

The disadvantage of our proposed algorithm is that it requires a large number of RSS measure samples to achieve good accuracy that can be a limiting factor in processing and storage capacity.

The contribution of the work is to evaluate the performance of the RFID system to offer the location aware service. More works are still needed to be done in order to make an ideal system. More factors should be taken into consideration in the future. These include the following:

- reduce the number of measurement samples required by using more efficient Bayesian filters and improved map generation algorithms;
- consider tests based on real measurement data to validate my model.

References

1. K. Finkenzeller, "RFID handbook: fundamentals and applications in contactless smart cards and identification" J. Wiley & Sons Ltd., New York, 2003.
2. J. Hightower and G. Borriello, "A Survey and Taxonomy of Location Sensing Systems for Ubiquitous Computing" IEEE Computer magazine, Vol. 34, No. 8, August 2001.
3. K. Pahlavan, L. Xinrong and J.P. Makela, "Indoor Geolocation Science and Technology", IEEE Communication Magazine, Vol. 40, No. 2, 2002.
4. E. Kaplan, "Understanding GPS: Principles and Applications", Artech House, Boston, 2005.
5. P. Bahl and V.N. Padmanabhan, "RADAR: An In-building RF-based User Location and Tracking System", Proc. IEEE Infocom, vol. 2, pp. 775–784, 2000.
6. J.D. Griffin, G.D. Durgin, A. Haldi and B. Kippelen, "Radio Link Budgets for 915 MHz RFID Antennas Placed on Various Objects", in WCNC Wireless Symposium .05.
7. D.M. Dobkin and S.M. Weigand, "UHF RFID and Tag Antenna Scattering, Part I: Experimental Results and Part II: Tag Array Scattering Theory", Microw. J. Theory Tech., 2006
8. P.R. Foster and R. Burberry, "Antenna problems in RFID systems, RFID Technology" IEE Colloquium RFID Tech., pp. 3/1–3/5, 1999.
9. T.S. Rappaport, "Wireless Communications: Principles and Practice", Prentice-Hall Inc., New Jersey, 2003.
10. "EPC Radio-Frequency Identity Protocols Class-1 Generation-2 UHF RFID Protocol for Communications at 860–960 MHz Version 1.0.9," EPC Global, Tech. Rep., January 2005.
11. R. Glidden et al., "Design of ultra-low-cost UHF RFID tags for supply chain applications", IEEE Communications Magazine, Vol. 42, No. 8, August 2004, pp. 140–151.
12. J.-P. Curty, N. Joehl, C. Dehollain and M.J. Declercq, "Design and Optimization of RFID system", Springer, Berlin, 2007.
13. D. Kim, M.A. Ingram and W.W. Smith, "Measurements of Small-Scale Fading and Path Loss for Long Range RF Tags," IEEE Trans. Antennas and Propagation, 51, 8, August 2003, 1740–1749.
14. J.G. Evans, R.A. Shober and S.A. Wilkus, "A Low-Cost Radio for an Electronic Price Label System" Bell Labs Tech. J., 22, 1996, 203–215.

15. K. Penttila, M. Keskilammi, L. Sydanheimo and M. Kivikoski, “Radar Cross-Section Analysis for Passive RFID Systems”, IEE Proc., Microw. Antennas Propag., 153, 1, 2006, 103–109.
16. http://www.cs.unc.edu/welch/kalman/
17. R.K. Mehra, “On the Identification of Variance and Adaptive Kalman Filtering” IEEE Trans. Automat, AC-15, 1970, 175–184.
18. http://www.mathworks.com/
19. J. Devore and R. Peck, “Statistics: The Exploration and Analysis of Data”, Duxbury Press, CA, 3 edition, 2004.

Adaptive Virtual Queue Random Early Detection in Satellite Networks

Do Jun Byun and John S. Baras

Abstract Networks with scarce bandwidth and high propagation delays cannot afford to have an unstable active queue management (AQM). In this paper, problems with applying existing AQMs to satellite networks are identified and solved. The first problem is oscillatory queuing, which is caused by high buffering due to performance enhancing proxy (PEP) in satellite networks where congestion control after the PEP buffering does not effectively control traffic senders. The second problem is global synchronization due to tail-drop nature of virtual queue-based AQMs. A new AQM method called adaptive virtual queue random early detection (AVQRED) is proposed to solve the problems, and it is validated using a realistic emulation environment and a mathematical model.

1 Introduction

Due to exponential increases in Internet traffic, active queue management (AQM) has been heavily studied by numerous researchers. However, little is known about AQM in satellite networks. A microscopic examination of queuing behavior in satellite networks is conducted to identify problems with applying existing AQM methods. A new AQM method is proposed to overcome the problems and it is validated with a realistic emulation environment and a mathematical model.

Internet Protocol (IP) over Satellite (IPoS) has been commercially available for the last few decades. Due to its high availability and mobility, IPoS has been attractive to areas where terrestrial services are not available as well as enterprises with scattered branch offices. One big barrier that IPoS has faced is its high propagation delay between earth stations and satellite. A typical round trip time (RTT) for a two-way geosynchronous satellite is around 600 ms.

D.J. Byun (✉)
Institute of Systems Research, University of Maryland at College Park,
e-mail: dbyun@hns.com

S. Powell, J.P. Shim (eds.), *Wireless Technology*, Lecture Notes in Electrical Engineering 44, DOI 10.1007/978-0-387-71787-6_4,

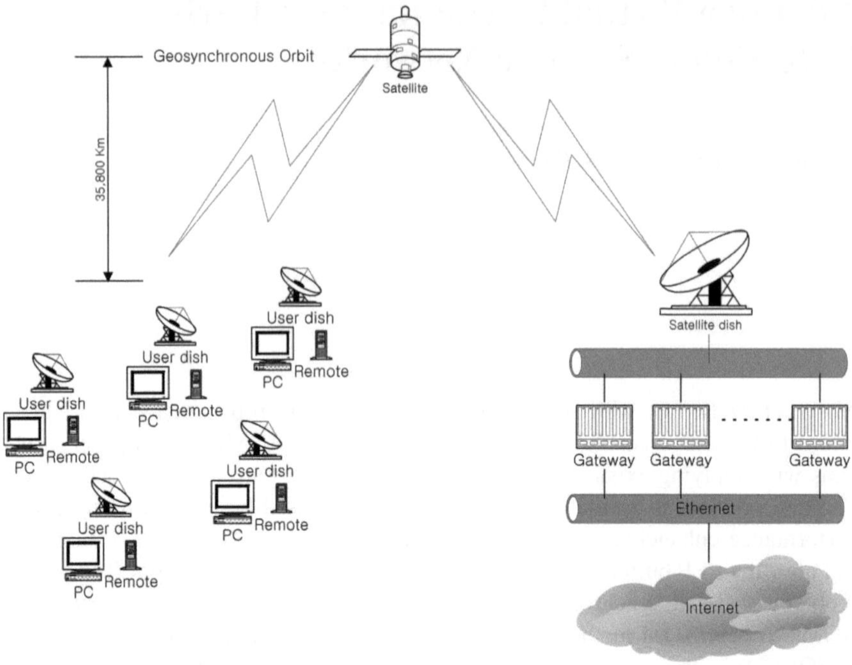

Fig. 1 IPoS system architecture

Figure 1 illustrates the system architecture of a typical two-way IPoS system where half of the 600 ms RTT occurs between the gateways and the satellite, the other half occurs between the remotes and the satellite. The biggest problem with such high propagation delay is the TCP performance. One aspect of the problem is the TCP slow start [1] phase where it takes a long time (RTT $\times \log_2 \times$SSTHRESH) to reach the maximum congestion window threshold (maximum rate at which the sender sends traffic) and the other aspect is that the maximum throughput of $65{,}535 \times 8/\text{RTT}$ is too low when the TCP Window Scale option is not supported. Even when the TCP Window Scale option is supported, unless all nodes support the option, fair bandwidth sharing becomes an issue. TCP Spoofing or performance enhancing proxy (PEP) [2] has been practiced by most of IPoS service providers to overcome this problem with TCP. For consistency, the term PEP will be used throughout this paper. The basic idea of PEP is to buffer at least one-round trip worth of data by locally acknowledging the data. Usually buffering only one-round trip worth of data is not enough because one has to account for queuing delays associated with congestion and inroute bandwidth allocations.

Active queue management (AQM) is an algorithm that detects and reacts to congestion to avoid queue overflows. There are generally two ways to react to congestion: signal congestion to traffic sources explicitly by setting explicit congestion notification (ECN) [3] bits or signal congestion to traffic sources implicitly by dropping packets. ECN is not used in our study due to the following reasons:

1. The problems that we are trying to solve are not due to packet drops between gateways and senders.
2. ECN marking after PEP (transmit queue in Fig. 3) may seem to avoid retransmissions over satellite and fix the queuing instability problem discussed later, but it is too late to enforce ECN bits when data are already acknowledged without ECN bits by PEP.

When applying AQM to satellite networks, the following need to be considered:

1. The source of congestion is different in satellite networks, i.e., in satellite networks, congestion arises mainly due to the satellite link capacity, not due to the processing capacity. Therefore, gateways in satellite networks become congested when the offered load is greater than the allowed transmit rate, whereas gateways in terrestrial networks often become congested when the offered load is greater than the processing capacity.
2. Monitoring and marking packets after PEP is not a good idea because it involves retransmissions over satellite.
3. Monitoring (with real-queue-based AQM) and marking packets before PEP is not a good idea because the receive queue will never be congested when the congestion bottleneck is the spacelink capacity, not the processing capacity. This is not true for virtual-queue-based [4] AQMs such as adaptive virtual queue (AVQ) [5].

To address the above concerns, a new virtual-queue-based AQM, Adaptive virtual queue random early detection (AVQRED), was previously proposed [6, 7] and validated with realistic emulations. In this chapter, we extend the study by constructing a mathematical model and further validate the solution by comparing the MATLAB results with the emulation results.

2 Overview of PEP

PEP enhances the TCP performance by locally acknowledging one+ round trip worth of TCP data at the gateway over terrestrial links. Although there can be many different flavors of PEP, the core idea of buffering up one+ round trip worth of TCP data remains the same.

Figure 2 illustrates the end-to-end PEP flows in a two-way satellite network. To better visualize PEP in a gateway, Fig. 3 is provided. PEP is drawn in a typical gateway structure where PEP processes packets after the receive queue and the transmit queue resides after PEP. In Fig. 3, the congested queue is the transmit queue as discussed in the previous section.

It is common that PEP is also implemented in remote terminals to gain higher upload speeds and to keep the implementation symmetric, but congestion avoidance

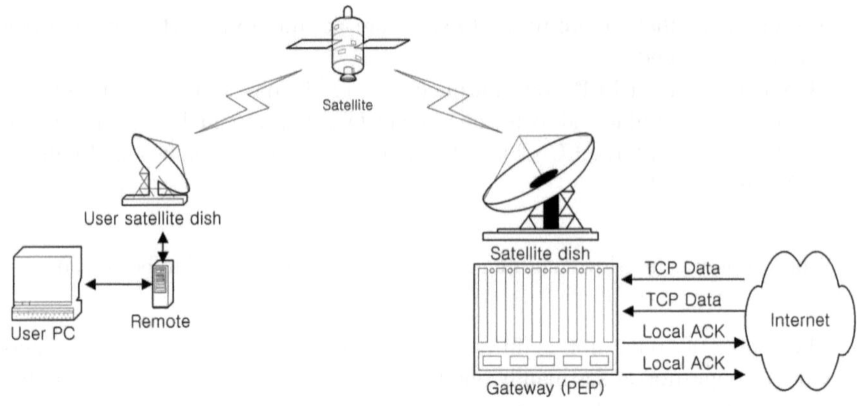

Fig. 2 PEP flows

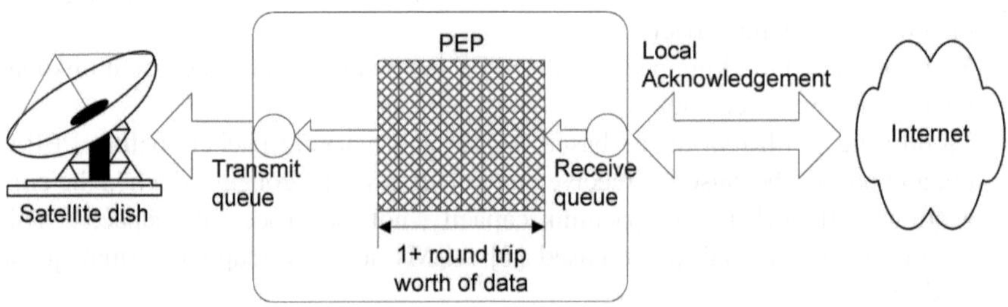

Fig. 3 Gateway with PEP

in the upload direction (from remote terminals to internet) is not discussed in this chapter as it involves different congestion paths.

3 Overview of AQM Methods

This section provides a high-level overview of two well-known AQM methods: random early detection [8] and adaptive virtual queue which will be compared with AVQRED via emulations and MATLAB.

3.1 RED

The RED [8] algorithm computes the marking probability when the weighted queue size falls between min_{th} and max_{th} parameters. The marking probability becomes higher as the weighted queue size gets closer to max_{th} (becomes 1 if it is greater than max_{th}), and it also becomes higher as the distance between each marking gets larger. Parameter tuning is required for

w_q and max_p. w_q controls the weighted average queue size which then determines how quickly the algorithm reacts to congestion. Reacting too quickly or too slowly may result in queuing instability. max_p is a scaling factor for the marking probability which also controls how quickly the algorithm reacts to congestion

```
Initialization:
avg = 0
count = -1
for each packet arrival
if the queue is nonempty
   avg =(1-wq)avg +wq.q
else
   m = f(time-q_time)
   avg = (1-wq)^m avg
if minth <= avg < maxth
   increment count
   calculate probability pa:
       pb = maxp(avg-minth)/(maxth-minth)
       pa = pb / (1-count.pb)
    with probability pa:
       mark the arriving packet
       count = 0
 else if maxth <= avg
    mark the arriving packet
    count = 0
 else count = -1
 when queue becomes empty
 q_time = time
```

RED algorithm

3.2 AVQ

Gibbens–Kelly virtual queue (GKVQ) [4] maintains a virtual queue whose service rate is the desired link utilization. When an incoming packet exceeds the virtual queue limit, it drops or marks the packet. Adaptive virtual queue (AVQ) maintains the same virtual queue whose capacity is dynamically adjusted. The virtual capacity is adjusted by adding the number of bytes that could have been serviced between the last and the current packets minus the bytes that were just received. Configured parameters are γ (target utilization), C (real capacity), and B (virtual queue limit).

```
At each packet arrival epoch do
/* Update Virtual Queue Size  */
```

```
VQ  = max (VQ - C'(t - s), 0)
If VQ + b > B
   Mark or drop packet in the real queue
else
   /* Update Virtual Queue Size */
   VQ = VQ + b
endif
/* Update Virtual Capacity */
C' = max (min (C'+α*γ*C*(t-s),C) - α*b, 0)
/* Update last packet arrival time */
s = t
```

```
Variables:

B   = buffer size
s   = arrival time of previous packet
t   = current time
b   = number of bytes in current packet
VQ  = number of bytes currently in the virtual queue
C'  = virtual capacity
C   = actual capacity
```

AVQ algorithm

4 Problems

The main objective we are trying to achieve is to avoid retransmissions over satellite, maintain queuing stability, and avoid global synchronization (consecutive packet drops) while preserving high link utilization. To avoid retransmissions over satellite, the option of dropping packets after PEP was not considered. Because real-queue-based AQMs such as RED can detect congestion only if it monitors the congested queue, RED monitoring is done in the transmit queue, while the packet marking is done in the receive queue to avoid retransmissions over satellite. Because virtual-queue-based AQMs such as AVQ can detect congestion regardless of the location of monitoring, both monitoring and marking are done in the receive queue to best synchronize packet marking and congestion detection by senders. Therefore, the following configurations are used throughout the emulations (Table 1).

Table 1 AQM Q and marking Q configuration

AQM method	Monitor Q	Marking Q
RED	Transmit queue	Receive queue
AVQ	Receive queue	Receive queue
AVQRED	Receive queue	Receive queue

4.1 Asynchronous Queuing Behavior

The problem with real-queue-based AQMs such as RED in satellite networks is synchronization between the monitored queue and the traffic senders. Synchronizing them is very difficult due to the high buffering that occurs between them, i.e., dropping a packet at the receive queue due to congestion in the transmit queue does not immediately reduce the congestion level of the transmit queue resulting in unwanted packet drops until the PEP buffers are all transmitted. These packet drops then result in less queue occupancy until senders' congestion windows evolve causing oscillatory queuing behavior. Figure 4 illustrates how an asynchronous queuing can occur. Note that the packets are consecutively dropped from T1 through T6 because the transmit queue is always occupied by the packets from the PEP layer. After PEP buffers are all used up, the transmit queue becomes almost empty and the PEP starts building up its buffers at T7. Until there are enough PEP buffers, the transmit queue does not drop packets at the receive queue causing oscillatory queuing behavior.

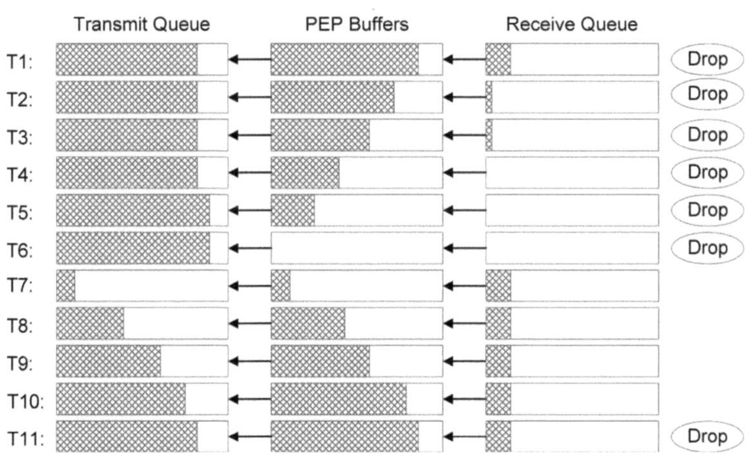

Fig. 4 Asynchronous queuing behavior

4.2 Global Synchronization

The problem with AVQ is global synchronization where consecutive packet drops occur due to its tail-drop nature of packet marking. When packets are dropped consecutively, multiple TCP connections will react to the drops simultaneously resulting in oscillatory link utilization among multiple TCP connections.

This problem is severe with RED due to its oscillatory queuing behavior described in the previous section. When the transmit queue congestion level and the senders congestion windows are not synchronized, the RED region will likely be exceeded resulting in tail-drop behavior.

5 Solution

A new AQM algorithm, adaptive virtual queue random early detection, is proposed to address the asynchronous queuing and the global synchronization problems.

```
for each packet arrival
/* Calculate virtual queue size */
δ <- curr_time - last_measure
if δ > 1
   /* Compute actual output rate in bps */
   tx_bytes <- bytes_transmitted
   output_rate <- (tx_bytes - prev_tx_bytes)* 8000 / δ
   prev_tx_bytes  <- tx_bytes

   /* Smoothen virtual capacity */
   v_capacity  <- α * output_rate + (1.0 - α) * v_capacity

   /* Update virtual capacity */
   v_capacity <- MAX (MIN (max_capacity,
                           v_capacity), min_capacity)

   /* # of bytes that could have been transmitted */
   serviced_bytes <- v_capacity / 1000 / 8 * δ

   if VQ  > serviced_bytes
      VQ <- VQ - serviced_bytes
   else
      VQ <- 0
      q_time <- curr_time

last_measure <- curr_time
q_size <- VQ / 1500

/* Feed VQ size to the RED algorithm */
if min_th < q_size < max_th
   count <- count  + 1
   p_b <- (q_size - min_th) /(max_th - min_th)
   p_a <- p_a / (1 - count * p_b)
   With probability p_a:
      Mark the arriving packet
      count <- 0
else if max_th <= q_size
    Mark the arriving packet
    count <- 0
else
```

```
count <-  -1
VQ    <- VQ + b
```

AVQRED Algorithm

The AVQRED algorithm constructs a virtual queue and feeds the virtual queue size to the RED algorithm instead of feeding the weighted average queue size to it. By doing so, AVQRED essentially moves the transmit queue to the receive queue and produces better synchronization between the transmit queue and the traffic sources. AVQRED reshapes the incoming traffic according to the desired link utilization because the RED algorithm reacts to the congestion level of the virtual queue which is serviced by the desired link utilization. The AVQRED algorithm above highlights the AVQRED parameters in bold. Note that w_q and max_p are no longer in the algorithm because their functionalities are replaced by the desired link utilization in AVQRED. α is a low-pass filter for the actual capacity calculation. *min_capacity* and *max_capacity* define the range of processing capacity. For satellite networks where processing capacity is greater than spacelink capacity, *min_capacity* should be equal to *max_capacity* and α can be any value.

5.1 Asynchronous Queuing Behavior

AVQRED solves the asynchronous queuing problem by both monitoring and marking at the receive queue. Monitoring and marking at the receive queue is possible because AVQRED constructs a virtual queue which can be placed anywhere.

5.2 Global Synchronization

AVQRED solves the global synchronization problem by preserving global synchronization avoidance of the RED algorithm. Emulation results are provided to illustrate this point.

6 Emulation Framework

The actual gateway software, IP Gateway, from *Hughes Network Systems* was used to evaluate the AQM methods. The three AQM methods were implemented according to Table 2. The emulation environment was constructed using two IP Gateways (one serves as the actual IP Gateway that faces the Internet and the other serves as the satellite terminals for *N* different users) and a traffic generator called *Spirent*. A high-level illustration of the gateway internal structures is shown in Fig. 3. Both server and client IP Gateways have the same PEP code and some modifications to the software were done to resolve address translation and routing issues created by the client IP Gateway. Details of the modifications are not discussed here as they are

Table 2 AQM settings

AQM	Parameters			
	min_{th}	max_{th}	w_q	max_p
RED 1	60	120	0.02	0.5
RED 2	60	120	0.05	0.7
RED 3	60	120	0.10	0.5
RED 4	60	120	0.10	0.7
	γ	B		
AVQ	100%	123,750 Bytes		
	min_{th}	max_{th}	*min_capacity*	*max_capacity*
AVQRED	60	120	20 Mbps	20 Mbps

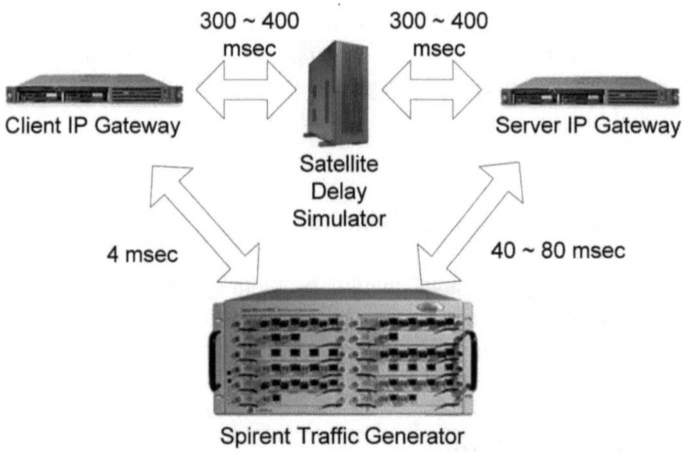

Fig. 5 Emulation flow

not relevant to the interest of this research. *Spirent* was used to best emulate real-life traffic characteristics.

Figure 5 illustrates the connectivity of the emulation setup. All links are lossless and 100 Mbps full duplex. A delay simulator was inserted between the two gateways to simulate satellite delays with uniform distribution between 300 and 400 ms each way. The round trip time (RTT) between the client IP Gateway and the *Spirent* is 4 ms, the RTT between the client IP Gateway and the server IP Gateway is 600–800 ms, and the RTT between the server IP Gateway and the Spirent is 40–80 ms resulting in an end-to-end RTT of 644–884 ms. Four hundred HTTP connections were generated between 200 clients and 60 servers with the following attributes.

1. At the startup, there are 20 new HTTP connections every 5 s with 5 s sleep time between each ramp up until 400 HTTP connections are established.
2. When a connection is closed, a new connection is created to fill the gap to maintain 400 HTTP connections.

3. Each web page contains 250–550 Kbytes of data with 10 s user think time.
4. The maximum download speed of each TCP connection is 5 Mbps.
5. Average birth and death rate of the connections is about 20 connections per second (approximately 5% of the total population).

6.1 Evaluation Methodologies

The following performance metrics were used for validation:

1. Link utilization – The purpose of this metric is to make sure that the proposed solution produces comparable link utilizations.
2. Queue size – The purpose of this metric is to compare queue size and queuing stability of each AQM method.
3. Packet drop – The purpose of this metric is to compare consecutive packet drops of each AQM method.

The measurements were taken after all 400 HTTP connections are established to best emulate a loaded scenario.

6.2 Parameter Settings

The following system parameters were used throughout the emulations:

- 20 Mbps downlink bandwidth (gateway to terminal direction).
- 1 Mbps uplink bandwidth (terminal to hub direction). This link is assumed to be non-congested link because the application is downlink-oriented web browsing.
- 5 ms transmit rate regulator latency in the server IP Gateway.
- Target transmit queuing delay of 33 ms.
- Average packet size is 1,400 bytes for the downlink direction.

For RED, there are four parameters to configure: min_{th}, max_{th}, max_p, and w_q. Sixty and 120 are configured for min_{th} and max_{th}, respectively. min_{th} is set slightly higher than $59(= 20\ \text{Mbps}/8/1{,}400 \times 0.033$ from the system parameters) to ensure full utilization of 20 Mbps bandwidth. max_{th} is set to at least twice min_{th} as [8] recommends. Several permutations of max_p and w_q were emulated as these parameters need to be fine-tuned according to traffic characteristics as shown in Table 2.

For AVQ, the target utilization, γ, is set to 100%, and the buffer size, B, is set to 123,750 bytes where $123{,}750 = 82{,}500\,(= 20\ \text{Mbps}/8 \times 0.033) + 82{,}500/2$. Half of the buffer required for 20 Mbps (82,500/2) is added to ensure full utilization. The α is set to an arbitrary number as our optimal virtual capacity is pre-determined.

For AVQRED, 60 and 120 are chosen for min_{th} and max_{th}, respectively with the same reason as RED; α is set to an arbitrary number as our optimal virtual capacity is pre-determined. Target utilization is set to 100% by setting *min_capacity* = *max_capacity* = 20 Mbps.

7 Emulation Results

Each of the AQM settings in Table 2 was emulated for 20 min with the traffic described in the previous section. The link utilization, queue size, and consecutive packet drop were measured once every 100 ms and the following subsections discuss the results for each of the measurement metrics. Due to the space limitation, only RED 4 (which had the best results amongs the RED settings), AVQ, and AVQRED are presented.

7.1 Link Utilization

As Figs. 6 and 7 and Table 3 show, the utilization of AVQRED is comparable with the utilization of RED. Although there is about 0.5% loss in the mean utilization, there is about 0.25% gain in the stability (the standard deviation).

Utilization loss and stability gain can be explained by the queuing behavior of RED and AVQRED which is discussed more in the next section. Basically, AVQRED maintains just enough data to fill up the 20 Mbps pipe whereas RED's utilization is oscillatory and unstable due to its asynchronous queuing behavior discussed in the previous sections. Furthermore, RED's high utilization and low stability indicate that it tends to accept more data than the gateway capacity.

Although AVQ's algorithm is similar to AVQRED's in terms of approximating the virtual capacity, its utilization is lower than AVQRED. This result is consis-

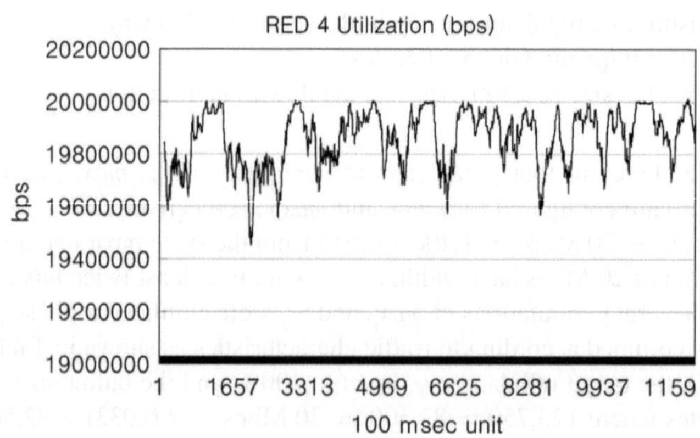

Fig. 6 RED 4 utilization

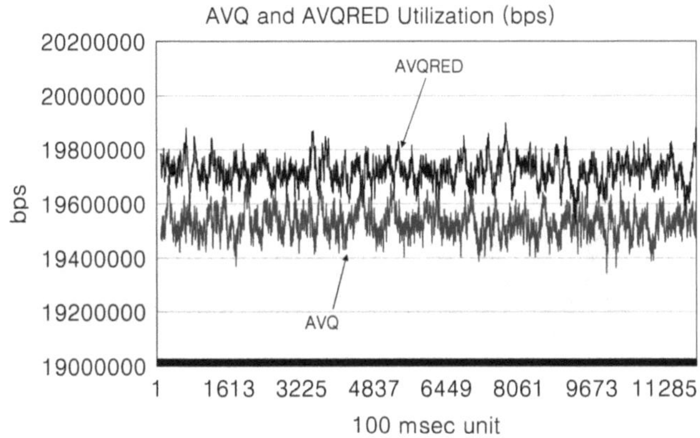

Fig. 7 AVQ and AVQRED utilization

Table 3 Link utilization mean and standard deviation

AQM	Mean (Mbps)	Standard deviation (Kbps)
RED 1	19.8	135
RED 2	19.8	117
RED 3	19.8	108
RED 4	19.8	107
AVQ	19.5	49
AVQRED	19.7	47

tent with the fact that AVQ has more consecutive packet drops because consecutive packet drops cause multiple senders to shrink their congestion windows synchronously resulting in lower link utilization.

7.2 *Queue Size*

Figures 8, 9, and 10 show the transmit queue size of RED, AVQ, and AVQRED. The queue size of RED is higher than AVQ and AVQRED because of its tendency to exceed the RED region (60–120) due to its oscillatory queuing behavior. To provide a better visualization of this point, Figs. 11 and 12 magnify Figs. 8 and 10 between 50th and 150th s (500th–1500th points according to the x-axis' scale).

This oscillatory queuing behavior is the asynchronous queuing behavior described earlier which is resulted from high PEP buffering between the transmit queue and the receive queue. Therefore, we can conclude that AVQRED and AVQ solve the asynchronous queuing problem by both monitoring and dropping at the receive queue. As discussed earlier, monitoring the receive queue with a real-queue-based AQM such as RED can not be done because the receive queue will never be

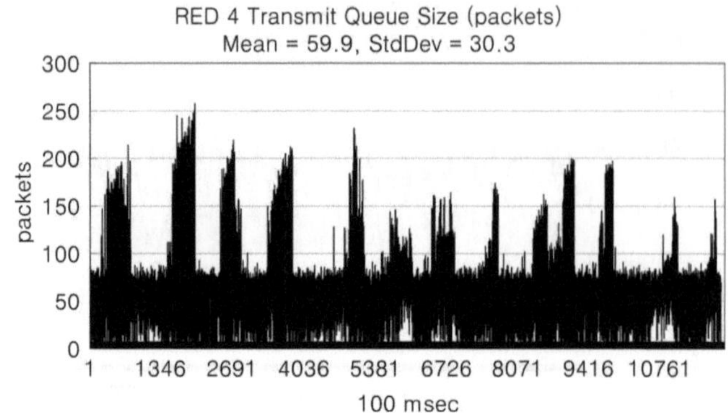

Fig. 8 RED 4 transmit queue size

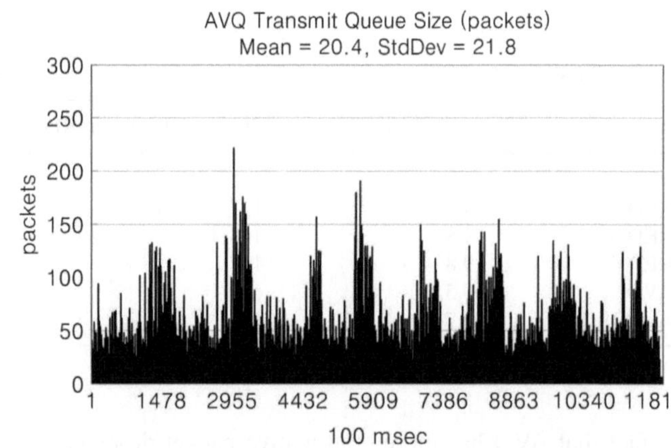

Fig. 9 AVQ transmit queue size

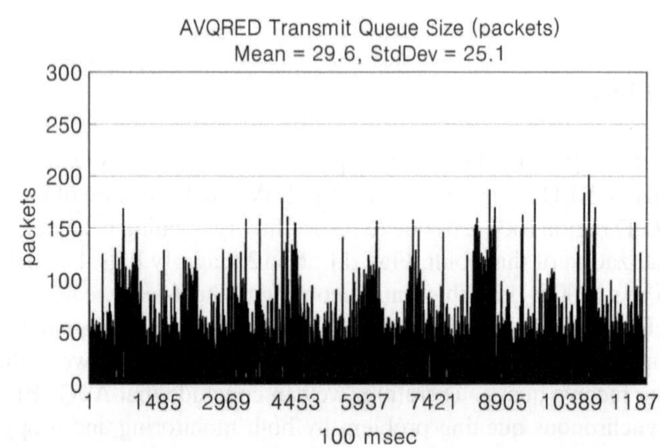

Fig. 10 AVQRED transmit queue size

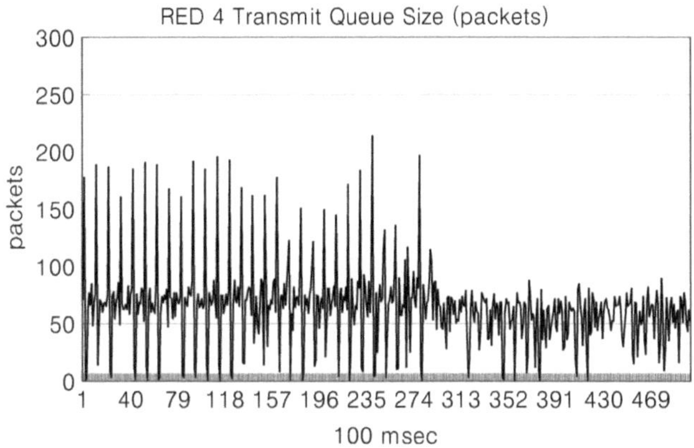

Fig. 11 RED 4 transmit queue size (50th–100th s)

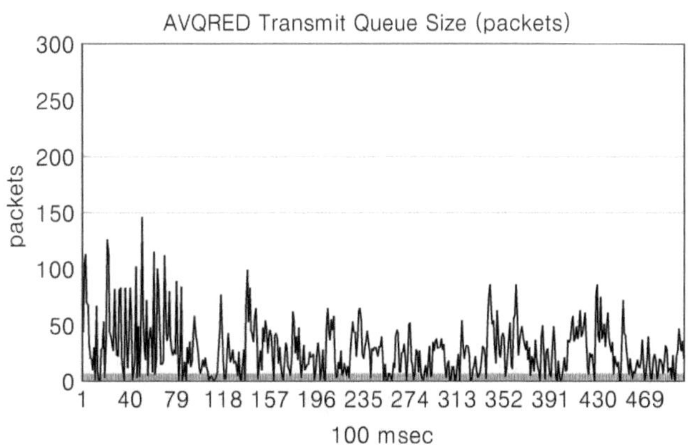

Fig. 12 AVQRED transmit queue size (50th–100th s)

congested when the bottleneck is the transmit queue by the spacelink bandwidth limitation.

7.3 Packet Drop

Because 20 min worth of the packet drop histogram is too long to present, only the first 1,000 packets are presented to show how packet drops are distributed. This illustration is valid because AVQRED has the least number of packet drops as shown in Table 4.

Figure 13 shows packet drops for the first 1,000 packets. Given that AVQRED has the least number of packet drops, having the least clustered packet drops proves that AVQRED has the least global synchronization level. RED packet drops are

Table 4 Total packet drops

AQM	Total packet drops
RED 1	486,932
RED 2	491,798
RED 3	492,025
RED 4	492,999
AVQ	484,639
AVQRED	484,582

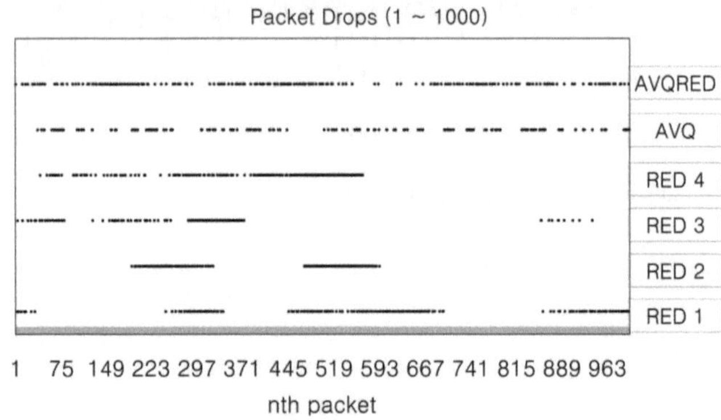

Fig. 13 Packet drops for 1st–1000th packets

more clustered than they should be due to the queuing oscillation discussed in the previous section.

From the data shown in this section, we can conclude that AVQRED solves the global synchronization problem of AVQ and RED by dropping packets more uniformly

8 Mathematical Model

This section provides a mathematical model for the queuing behavior to validate the asynchronous queuing problem and our solution, AVQRED. The global synchronization problem is not validated mathematically due to the space limitation. However, the same model can be used to show the marking behavior by analyzing the standard deviation of the marking probability:

$$\frac{d\lambda}{dt} = (1 - p(t)) \times \frac{m}{R^2} - p(t) \times \frac{\lambda^2}{2m} \tag{1}$$

$$\frac{dq}{dt} = -\mu + (1 - p(t - d(t))) \times \omega(t - d(t)) \times \lambda(t - d(t)) \tag{2}$$

$$\frac{dv}{dt} = -\tau \times \mu + (1 - p(t)) \times \omega(t) \times \lambda(t). \tag{3}$$

$$\frac{dw}{dt} = \frac{\log(1-B)}{\delta} \times w(t) - \frac{\log(1-B)}{\delta} \times q(t) \tag{4}$$

$$\frac{dp}{dt} = \begin{cases} -1.0 \times p(t) & \text{if } w(or\ v) < min_{\text{th}} \\ \frac{p_{max}}{(max_{\text{th}} - min_{\text{th}})} \times \frac{dq}{dt} & \text{if } min_{\text{th}} \le w \le max_{\text{th}} \text{ } for\ RED \\ \frac{p_{max}}{(max_{\text{th}} - min_{\text{th}})} \times \frac{dv}{dt} & \text{if } min_{\text{th}} \le v \le max_{\text{th}} \text{ } for \text{ AVQRED} \\ 1.0 - p(t) & else \end{cases} \tag{5}$$

Equation (1) is the ODE of the arrival rate of the offered load where R is the RTT between the gateway and the Internet hosts and m is the number of TCP connections. Ref. [9] has the details on how it is mathematically derived. In our MATLAB experimentation, R was scaled down by 1/10 due to our time unit conversion from 1 s to 100 ms.

Equation (2) is the ODE of the transmit queue size where μ is the service rate (20 Mbps with 1,400 bytes per packet and 100 ms time unit), $p(t)$ is the marking probability, $d(t)$ is the Fourier series of the PEP buffering delays, and $\omega(t)$ is the Fourier series of the offered load variation. The ODE is derived from the Lindley equation and the delay factor was added to it to capture the PEP buffering effect. To best resemble our traffic model used for the emulations, Fourier series with 500 actual data points were used. For $d(t)$, the data points are the average duration that each PEP packet resides in the buffer during a 100 ms measurement period. For $\omega(t)$, the data points are the mean offered load to the actual offered load ratio. All the data points were measured without AQM and bottlenecking transmit queue to avoid any feedback effects caused by AQM.

Equation (3) is the ODE of the AVQ size where τ is the target utilization. Note that it is similar to (2) except that it does not have the PEP buffering delays.

Equation (4) is the ODE of the weighted average transmit queue size from [10]. B is w_q and δ is the smallest time unit of our ODE approximation which is 1 ms (= 0.01 of 100 ms).

Equation (5) is the ODE of the marking probability which is just the first derivative of $p(t)$ when the respective queue size (q or v) falls between min_{th} and max_{th}. For AVQ, this needs to be changed slightly,
i.e., $-1.0 \times p(t)$ if $v < (max_{th} - min_{th})/2$, and $1.0 - p(t)$, otherwise.

To validate the asynchronous queuing problem, the above ODEs were fed to MATLAB and the transmit queue size (2) was examined for both RED (with RED 4 parameters in Table 2) and AVQRED. As Fig. 14 shows, RED has the same oscillatory queuing behavior as the one that Fig. 8 shows. The mean is slightly lower than the emulation because the ODEs did not account for the 5 ms queuing latency caused by the output rate regulation. The standard deviation is slightly higher than the emulation because the Fourier series for the PEP buffering delays was

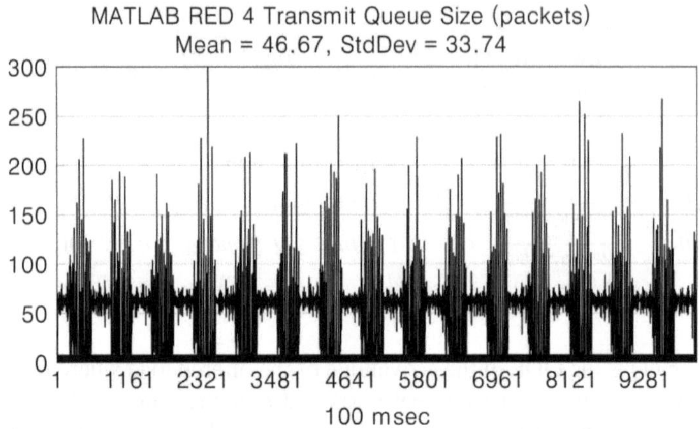

Fig. 14 RED 4 transmit queue size (MATLAB)

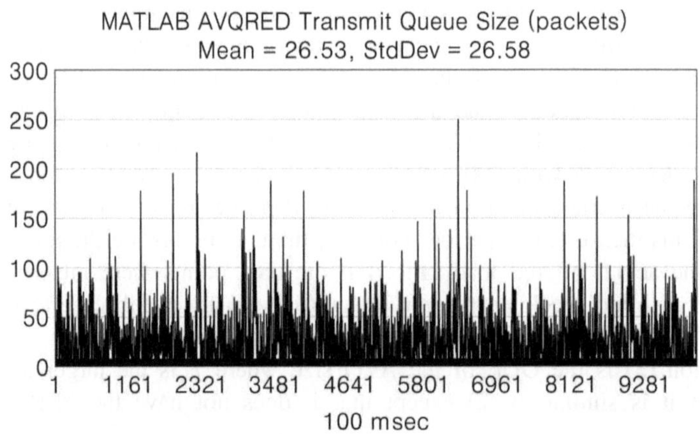

Fig. 15 AVQRED transmit queue size (MATLAB)

approximated using only 500 data points which resulted in more frequent and regular oscillations.

As Fig. 15 shows, AVQRED fixes the oscillatory queuing behavior. The mean and standard deviation are slightly different from the emulation because of the 5 ms latency and the relatively small Fourier sample space.

To summarize and compare the improvement percent of mean and standard deviation, Fig. 16 is provided. Figure 16 depicts that the MATLAB results concur with the emulation results. As stated earlier, the small discrepancies between emulation and MATLAB are from the 5 ms rate regulator latency and the small Fourier sample space.

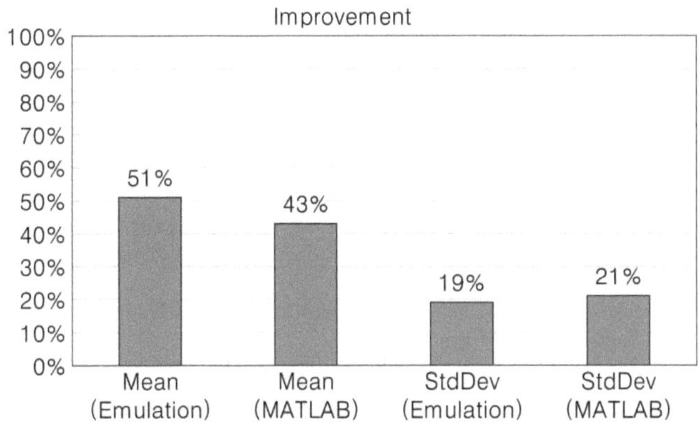

Fig. 16 Transmit queue size improvement by AVQRED

9 Conclusion and Future Work

In an effort to improve the gateway performance of satellite networks, AQM was applied to satellite networks. This study found that applying existing AQMs such as RED and AVQ has unwanted side effects: asynchronous queuing and global synchronization.

Emulations were conducted to validate the problems and the solution. The emulation environment was constructed with the real gateway software used in *Hughes Network Systems* and a traffic generator called *Spirent*.

A mathematical model was constructed to provide intuitive illustrations of the problems and the solution. The model was fed to MATLAB and the results concurred with the emulation results.

Because scarce bandwidth resources such as satellite require a good QoS handling, our future work will impose QoS as another measurement metric and involve enhancing the AVQRED algorithm to account for QoS related requirements.

References

1. W. Stevens, "TCP slow start, congestion avoidance, fast retransmit, and fast recovery algorithms," RFC2001, January 1997.
2. J. Border, M. Kojo, J. Griner, G. Montenegro and Z. Shelby, "Performance enhancing proxies intended to mitigate link-related degradations," RFC3135, June 2001.
3. K. Ramakrishnan and S. Floyd, "A proposal to add explicit congestion notification (ECN) to IP," RFC 2481, January 1999.
4. R.J. Gibbens and F.P. Kelly, "Distributed connection acceptance control for a connectionless network," in *Proceedings of the 16th Intl. Teletraffic Congress*, June 1999.
5. S. Kunniyur and R. Srikant, "Analysis and design of an adaptive virtual (AVQ) algorithm for active queue management," in *Proceedings of ACM/SIGCOMM*, August 2001.
6. D. Byun and J. Baras, "Adaptive virtual queue random early detection in satellite networks," in *Proceedings of Wireless Telecommunication Symposium*, April 26–28, 2007.

7. D. Byun and J. Baras, "A new rate-based active queue management: adaptive virtual queue RED," in *Proceedings of the Fifth Annual Conference on Communication Networks and Services Research*, New Brunswick, Canada, May 14–17, 2007.
8. S. Floyd and V. Jacobson, "Random early detection gateways in congestion avoidance," *IEEE/ACM Trans. Netw.*, 1(3), 397–413, 1993.
9. P. Kuusela, P. Lassila, J. Virtamo and P. Key, "Modeling RED with idealized TCP sources," http://research.microsoft.com/~peterkey/Papers/ ifipredtcp.pdf, 2001.
10. V. Misra, V. Gong and D. Towsley, "A fluid-based analysis of a network of AQM routers supporting TCP flows with an application to RED," in *Proceedings of ACM SIGCOMM*, August 2000, pp. 151–160.
11. P. Lassila and J. Virtamo, "Modeling the dynamics of the RED algorithm," in *Proceedings of QofIS'00*, September 2000, pp. 28–42.
12. J. Padhye, V. Firoiu, D. Towsley and J. Kurose. "Modeling TCP throughput: A simple model and its empirical validation," in *Proceedings of ACM/SIGCOMM*, 1998.
13. W. Stevens. TCP/IP Illustrated, Vol. 1, *The Protocols.* Addison-Wesley, Boston, 1994.
14. W. Stevens, "TCP congestion control," RFC2581, Apr 1999.
15. K. Ramakrishnan, S. Floyd and D. Black, "The addition of explicit congestion notification (ECN) to IP," RFC3168, September 2001.
16. M. Karaliopoulos, R. Tafazolli and B. Evans, "Proxy-assisted TCP maximum receive window control in split-TCP capable GEO satellite networks," in *Proceedings of American Institute of Aeronautics and Astronautics*, May 9–12, 2004.

News Corporation: Facing the Wireless World of the 21st Century

Vassiliki Th. Cossiavelou and Michael R. Bartolacci

Abstract A new generation of media consumers has arisen this century that is becoming more and more accustomed to content that is delivered via a wireless network. News Corporation, a global media industry giant, is taking on this challenge of providing sports, news, and other content to these consumers in a variety of ways. This chapter looks at the investments and strategies employed by News Corporation in delivering content via a wireless means for both today and in the future.

1 Introduction

Tremendous changes in the world are occurring at a pace never experienced by our ancestors. These changes that will reshape the human experience in the 21st century will be covered by the media industry. Wireless technologies will play a key role for the media industry as these changes impact societies and countries. The new generation of media consumers that has arisen in the last decade or so demands content delivered when they desire and in the format they desire. The emerging wireless-driven generation is the force behind the knowledge revolution in the media industry. The media industry is relying more and more on state-of-the-art wireless technologies. News Corporation is preparing to face this brave new world with innovations in its use of wireless technologies, in particular its use of satellites and the porting of content to emerging wireless platforms. It is also developing strategic alliances that revolve around these wireless technologies worldwide including the emerging Asian markets. Its strategic business operations that are focusing on wireless platforms allow it to be a role model for other media organizations to follow.

V.Th. Cossiavelou (✉)
Communication Secretary A', Ph.D. Candidate, Aegean University, Greece
e-mail: v.cossiavelou@ct.aegean.gr

S. Powell, J.P. Shim (eds.), *Wireless Technology*, Lecture Notes in Electrical Engineering 44, DOI 10.1007/978-0-387-71787-6_5,

The third communications strategy and policy paradigm, which followed the public service media policy (1945–1980/1990) [1], is a perfect fit for News Corporation (News Corp). This strategy and policy paradigm depicts a diversified and global scope for communications/media companies such as News Corp. News Corp has more than two centuries of media "know how." It is one of the top three media giants of the world and had total assets as of June 30, 2006, of approximately $57 billion with total annual revenues of approximately $25 billion [2]. News Corp is a diversified international media and entertainment company with operations in eight industry segments: film entertainment; traditional broadcast television; cable network programming, direct broadcast satellite television, magazines and inserts, newspapers, book publishing, and other media. The activities of News Corp are conducted principally in the United States, continental Europe, the United Kingdom, Australia, Asia, and the Pacific Basin [3]. News Corp competes in the entertainment business with other diversified international entertainment companies such as Time Warner, Viacom, Sony Corporation, The Walt Disney Company, and NBC Universal. As new technologies for delivering content and services evolve, News Corp is pursuing opportunities to distribute content to consumers through various media outlets including the Internet, mobile devices, video-on-demand, interactive television, and video games. As media conglomerates consolidate even further in the current global business environment, News Corp has developed a business strategy that includes a significant reliance on wireless technologies to survive and prosper as it heads for the middle of this century.

The emerging platform of the wireless "third screen," after the television and personal computer, has an advantage in the content market for media conglomerates such as News Corp. In 2004 almost 25% of the world's population consisted of mobile phone/device subscribers, eclipsing fixed line telephone, and Internet users for the first time. In 2009 the total number of mobile phone/device subscribers is expected to reach the 2.5 billion. If this growth continues, that number could reach 9.7 billion by 2054 which would equate to more than one mobile receiver per user [4]. This growth potential represents significant opportunities for News Corp for all major types of mobile content including news/information, entertainment, marketing, and promotion. Challenges such as digital rights management and interoperability issues exist ahead, but it is positioned to realize such growth. The key to such growth appears to hinge upon a seamless, platform-neutral strategy with wireless playing a vital role.

2 News Corporation Worldwide Wireless Interests

The management of media conglomerates has entered into a period of consolidation. The strategy of News Corp has been to follow this trend as well. News Corp's interests in the wireless field include (in terms of channel infrastructure in satellite television)

1) BSkyB, United Kingdom (37.9% holding);
2) DirecTV Group, North and South America (36.8% holding) which is comprised of DirecTV (the largest USA satellite TV provider);
   DirecTV Latin America, Sky Brazil (co-owned with Globopar), Sky Mexico;
3) Foxtel, Australia (25% holding);
4) Sky Italia, Italian satellite TV service;
5) STAR TV, an Asian satellite TV service;
6) Tata Sky, an Indian DTH TV service (20% holding) [2].

News Corp's wireless services also include podcasting sports, news and entertainment headlines and videos via Sky Sports and Sky News, and weather forecasts from the Sky News Weather, exclusive entertainment field interviews, reviews and previews from Sky One and Sky Movies. Sky Mobile TV, another News Corp entity, is offering over 19 channels direct to Vodafone 3G (third-generation mobile phone) users. On November 1, 2005, Vodafone became the first UK mobile network to deliver Sky Mobile TV for Sky Sports News and Sky News. The programs are streamed to each mobile handset and therefore are not as taxing on a device's memory as downloaded content. During 2005, FOX News became the first news channel available to mobile phone users in the US, appearing live on Sprint phones 24 h a day, 7 days a week.

> News Corp's Sky Mobile TV is currently offering over 19 channels direct to users via Vodafone's 3G subscriber base under such wireless content channels as Vodafone Live!, Sky Sports News and Sky News. On November 1, 2005, Vodafone became the first U.K. mobile network to deliver Sky Mobile TV. The programs are streamed to mobile subscribers and since they are not downloaded to handsets, they do not consume large amounts of local memory.

3 Satellite TV and Satellite Mobile TV

The goals for satellite television broadcasters of expanding subscription rates/market shares and efficient management while promoting cultural diversity belong to different patterns of inter- and intra-industry competition. [5, 6] In the US, all four satellite television competitors that existed before 1994 (DirecTV, USSB, Prime Star, and DISH Network) *today* are merged with DirecTV in essence and looking worldwide control about the 58% of the total amount of subscribers in the US, Japan, England, and France [7]. To accomplish these goals, the efficient management of resources and a promotion of cultural diversity must also take place as a growing global footprint through market-share acquisition. Owen and Wildman [5] argue that these two are critical success factors within the various patterns of inter- and intra-industry competition. In 2000, News Corporation announced plans to consolidate its worldwide satellite platforms and certain related assets under one umbrella entity [8].

News Corp later named this umbrella entity Sky Global Networks which comprised the equity interests of satellite distribution platforms including British Sky Broadcasting, STAR, Stream, Sky Brazil, Sky Mexico, Sky Multi-Country Partners, and SKY PerfecTV!, as well as its equity interests in NDS Group and TV Guide [8]. Sky Global Networks has approximately a 21% equity interest in Gemstar-TV Guide International, Inc. In July 2000, Gemstar International Group Limited and TV Guide, Inc. announced completion of their merger. Gemstar′s technology and intellectual property are licensed to 18 major companies in the telecommunications industry. As of a June 30, 2005, Gemstar-TV Guide had interactive program guide patent license agreements or interactive program guide product agreements with satellite and cable distributors representing 36 million digital subscribers, a 23% increase over the previous year [9]. Such a significant presence in the global satellite content business clearly shows that News Corp has staked its future in the wireless global arena.

The DIRECTV Group also launched two new satellites to have the capability to deliver high definition local channels to about half of American households. Two additional satellites have been launched in 2007 to provide the capacity to deliver local HD service to every home in America. This last point is a noteworthy one if media content ported to the wireless environment is to be ultra-successful.

4 Sports and Entertainment: Fuelling the Wireless ontent

It is obvious that News Corp realizes the importance that sports and entertainment content have within the wireless global media world. The strategic challenge for media companies of delivering mobile Internet hinges on the fact that three options are available: syndication to distribute existing content to new intermediaries or third party portals, developing a portal that builds on their media brands, and using an MVNO (mobile virtual network operator) [9]. For all four of the types of mobile content applications for media industries, i.e., information and entertainment, marketing and promotion, communications, and transactions, the value chain framework becomes enriched. In January 2003, News Corp and Cablevision Systems Corporation, one of the USA.'s leading entertainment and telecommunications companies, reaffirmed their efforts on a partnership for the delivery of sports and entertainment content. News Corporation's STATS Inc., the leading sports information and statistical analysis company in the US, recently moved to porting its content across multiple media platforms including the wireless environment. In 2005, News Corp also completed its reorganization in the US with the acquisition of the remaining 18% interest in Fox Entertainment Group and took full ownership of Fox Sports Networks Ohio and Fox Sports Networks Florida. Current plans include consolidation of direct broadcast satellite operations for these and other related sports entities into single business units. It is obvious that News Corp realizes the importance that sports and entertainment content have within the wireless media world.

In 2005 News Corp also completed its reorganization within the US with the acquisition of the remaining 18% interest in Fox Entertainment Group and took full ownership of Fox Sports Networks Ohio and Fox Sports Networks Florida. In July 2005, News Corporation also formed a new Internet unit, the Fox Interactive Media (FIM) to consolidate News Corp's US web properties and to leverage its strength in entertainment, news, and sports brands across the Internet. FIM also allowed it to offer a richer online experience to its millions of users. The focus of FIM was on creating a robust, broadband experience that leveraged the vast current and archived video assets of the company while also building an integrated hub of dynamic content with multiple points of entry, seamless navigation, and the ability to customize and personalize.

5 The Management of the Wireless Omnipresence to Consumers

The implementation of an integrated wireless media experience strategy allowed News Corp in May 2000 to create an international wireless joint venture with OmniSky. It invested $60 million to explore and expand international opportunities for the deployment of high-speed wireless Internet access, content, and e-commerce. OmniSky is a leading provider of comprehensive branded wireless Internet services for handheld mobile devices. At that time the prediction was for over 1 billion cellular communication devices being used globally by 2003 and for 40% or more of those devices to be data enabled for news, sports and entertainment, finance, shopping, travel, and weather.

News Corp's Fox Mobile Entertainment group got its start with the very successful "American Idol" television program text voting which generated nearly 65 million text messages during the 2007 season, up from 12,000 messages in the first season of 2001. In a long list of firsts, the company also invented the Mobisodes Series category. This category started with the wirelessly -delivered "24: Conspiracy" series, the first made-for-mobile program to be Emmy-nominated, and launched the first ad-sponsored video series, "Prison Break: Proof of Innocence." In February 2006, Fox Mobile Entertainment launched the first media-backed cross-carrier mobile entertainment service for consumers, Mobizzo. This is a comprehensive, new destination for mobile content from across News Corp's worldwide divisions, as well as from other media companies and up-and-coming talent. The Fox Mobile Entertainment group already has licensing and distribution deals for Fox's top brands with the world's most powerful carriers that reach nearly 2 billion mobile phone subscribers worldwide, including 200 million in the US. Mobizzo is an innovative "virtual place" for fans to find their favourite entertainment shows and theoretically also enjoy the benefits of competitive pricing with excellent customer service. A la carte offerings ranging from $1.99 to $2.49 and monthly subscription plans averaging $5.99 per month were available at the time of Mobizzo's introduction. Fox Mobile Entertainment, by the beginning of 2006, had distributed four series and 200 Mobisodes that had been seen in 25 countries and translated into six languages [4].

6 The Global Media Marketplace and Wireless for News Corp

The forecasted new "thumb" generation of 9.1 billion people worldwide in 2050 [10] represents potential media consumers that will have arisen demanding content delivered at their chosen time and manner. This emerging generation is forming virtual communities and is fuelling the technological revolution where secure interoperable platforms are already being dominated by wireless technologies. Analysts forecast that future media brands will be based on a seamless, platform-neutral, "bricks-and-clicks" marketing, and distribution strategy.

News Corp, through its subsidiary News Digital Systems (NDS), was able to manage its digital rights content very well in the hard-wired access environments. The "coming of age" of the integrated wireless media experience led to the creation of a joint venture with OmniSky for the secure delivery and management of content in the wireless environment.

In September 2006 the Global Mobile Entertainment Company which was born as a News Corp and VeriSign joint venture became the world's largest provider in delivery of mobile entertainment. VeriSign operates intelligent infrastructure services that enable and protect interactions across voice and data networks. News Corp will pay approximately "188 million for a controlling interest in VeriSign's wholly owned Jamba subsidiary and will combine it with Fox Mobile Entertainment assets. The new company will merge the most technologically advanced content delivery platform in the category with the market-leading mobile content production and delivery capabilities. The joint venture will serve 30 international regions with a potential reach of more than a billion mobile subscribers. It will also become the largest customer for VeriSign's Digital Content Services (DCS) group, which specializes in providing intelligent infrastructure and connectivity solutions to enable the delivery of rich content over mobile and broadband networks. Under the agreement, Jamba will soon release its first two products and offerings as a new entity, including MySpace Mobile Store. The move of MySpace, with more than 74 million users worldwide, into the wireless world complements other offerings aimed at a younger, wireless-savvy demographic.

7 News Corporation's Contribution to Secure Wireless Delivery

In February 2000, News Corporation subsidiary NDS (News Digital Systems) in the UK, which also has offices in 10 countries across 5 continents, joined Eastman Kodak's PictureVision subsidiary in a joint venture. NDS's major product is the VideoGuard CAS, which is used on most of News Corporation's digital satellite TV systems and on many non-News Corp systems, for offering open and flexible software solutions for the secure delivery of entertainment and information to televisions and PCs.

In July 2000, News Corporation, concerned about the future of copyright protection, addressed the Centre National Policy in Washington in order to stimulate legislation to protect digital content. As the move into the wireless arena grew, News Corp and its subsidiaries realized that digital content piracy would be a major obstacle to overcome and initially, a thorn in its side. News Corp looked forward to working with Intel and other groups to solve the problem of digital piracy and applauded the Joint Statement of Principles agreement between AOL-Time Warner and Intel on copyright protection. In 2002 the Advanced Television Systems Committee (ATSC), an international non-profit membership organization, developed voluntary standards for all advanced television systems and voted affirmatively on News Corp's proposal to label DTV broadcasts with the new voluntary standard "broadcast flag." This is an important first step in enabling digital broadcasters to prevent unauthorized redistribution of their content on the Internet.

8 Conclusions

News Corporation is leading the way into the wireless media world of the 21st century. Only time will tell what this world will be like by mid-century, but through its joint ventures, its strategic business planning, and its willingness to embrace the emerging wireless technologies, News Corp will certainly be a major player and a role model for the media industry in wireless sector. Its strategy of globalization and personalization of the wireless media experience has certainly given it the ability to leverage its vast media content repository and develop new content directly suited for the wireless platforms; thus preparing itself for serving the eventual "glocal" markets of 2050. The sports and entertainment sectors are key parts of News Corp's wireless strategy and represent a major source of revenue potential in the future.

References

1. J. van Cuilenburg and D. McQuail (2003) Media Policy Paradigm Shifts, Towards a New Communications Policy Paradigm, European Journal of Communication, 18(2), 181–207
2. Annual Report "Imagine the future" (30-6-2006), Retrieved on http://www.newscorp.com/Report2006/AnnualReport2006/HTML2/news_corp_ar2006_0115.htm
3. Full Description News Corp NWSa (NYSE) (3/1/2007), Retrieved on http://stocks.us.reuters.com/stocks/fullDescription.asp?symbol=NWSa.N&WTmodLoc=BizArt-L2-MarketView-3

4. S.M. Chan-Olmsted (2006) "Content Development for the Third Screen: The Business and Strategy of Mobile Content and Applications in the United States", The International Journal on Media Management, 8(2), 51–59
5. B. Owen and S. Wildman, Video Economics, Harvard University Press, Cambridge, 1992
6. Management and Economic Implications of Bundling and Block Booking of Television and Cable Programming, Picard, Robert G., Paper presented at the Annual Meeting of the Association for Education in Journalism and Mass Communication (72nd, Washington, DC, August 10–13, 1989
7. S. Seunghye (2005) Interindustry and Intraindustry Competition in Satellite Broadcasting: A Comparative Case Study on the United States, Japan, England, and France, Journal of Media Economics, 18(3), 167–182
8. A. Butcher, (14/2/2000), "Statement Regarding Satellite Platforms" New York, www.newscorp.com/news/news_101.html
9. F. Valerine (2002) Competitive Strategy for Media Companies In The Mobile Internet, Schmalenbach Business Review: ZFBF; Oct 2002; 54, 4; ABI/INFORM Global, pp. 351–371
10. www.un.org/News/Press/docs/2005/pop918.doc.htm United Nations (24-2-2005) "World Population To Increase By 2.6 Billion Over Next 45 Years, With All Growth Occurring In Less Developed Regions Numbers to Rise from Present 6.5 Billion, Hitting 9.1 Billion by 2050", New York

Delay Effect on Conversational Quality in Telecommunication Networks: Do We Mind?

Jan Holub and Ondrej Tomiska

Abstract Transmission delays are unwanted effects of every day telecommunication systems. These effects influence the overall quality of a telephony connection as perceived by end users. This chapter summarizes known subjective experiments testing the influence of delays on the quality perceived by end users. In order to find a possible explanation for the differences in the results of those experiments a new conversational test was conducted. The new subjective tests are also presented and compared to already existing experiment results.

1 Introduction

Speech transmission quality in telecommunication networks is affected by impairments like transmission delay, echo, noise, speech coding distortions, temporal, and amplitude clipping. The overall quality can be evaluated and expressed in terms of a mean opinion score (MOS) covering the range from 1 (bad quality) to 5 (excellent quality) [6].

Presently only impairments affecting the perceived quality of a transmitted speech signal are evaluated. This is done either subjectively by means of listening-only tests [6] or objectively using suitable algorithms. Some impairments, however, like echo or delay can only be assessed by conversational tests. Since objective and perceptually motivated measurement techniques are not yet available conversational tests must be conducted subjectively. In case of listening-only tests, the results are called MOS-LQ while for conversational tests the results are named MOS-CQ [7]. In both cases, an additional letter is added to indicate if the results were obtained

J. Holub (✉)
Dept. of Measurement K13138, FEE CTU Prague, Technicka 2, CZ 166 27 Prague 6, Czech Republic
e-mail: holubjan@fel.cvut.cz

S. Powell, J.P. Shim (eds.), *Wireless Technology*, Lecture Notes in Electrical Engineering 44, DOI 10.1007/978-0-387-71787-6_6,

subjectively (S) or objectively (O) [7]. For example, MOS-CQS denotes a subjective conversational test result.

In this chapter, we address the effect on the perceptual impact of various transmission delays in echo-free and also echo-present telephony situations. Several subjective tests evaluating perceptual annoyance of transmission delay already exist. In the next chapters some of them will be described and compared with each other. Also new results of tests performed by the authors of this chapter will be presented.

1.1 Delay in Modern Telecommunication Networks

The causes for transmission delays are numerous of which only a short overview shall be provided in this document. A more detailed list of delay causes can be found in [2].

Sending terminal factors:

- Speech coding delay
- Packetization
- Serialization

Network factors:

- xDSL transmission/processing delay or
- Radio link delay
- Backbone propagation delay
- Queuing delay
- VoIP gateway delay

Receiving terminal factors:

- Playout buffer delay
- Speech decoding delay

1.2 Conversational Tests Versus Listening-only Tests

Conducting conversational tests according, e.g., to [6] are more complex than listening-only tests. This is because conversational tests require special premises with at least two acoustically decoupled environments with defined acoustical parameters like background noise and reverberation time. Also a network simulator allowing at least for changing transmission delays is necessary, since a live network must be simulated, while in listening-only tests the subjects are presented with recordings of speech signals.

Conversational tests can therefore also only be run one at a time, whereas in listening-only tests many listeners may evaluate a telephony situation in parallel.

1.3 Conversational Interactivity Versus Delay Annoyance – Test Scenarios

The perception of delay strongly depends on the conversational situation. If the purpose of a conversational test is to assess the network quality as it is perceived by an end user in a real telephony situation the test scenarios should be close to that situation. This is because conversational situations are strongly influenced by the degree of interaction between the participants. For example, in stressful situations where lots of information must be transmitted in very short time the subjects become more sensitive to delay effects as they more often may interrupt each other unwillingly (double-talk).

Different conversational tests evoke different conversational interactivity. This parameter is defined and measured by multiple ways [10, 2]. According to [2] it is sufficient to evaluate the number of role (listener-talker) swaps per minute between the conversation participants. This parameter is denoted as SAR (speaker alternation rate, min^{-1}).

In some chapters like [10, 11, 2] authors propose various methods (conversational scenarios) forcing higher or lower degrees of interactivity. For example, Kitawaki in 1991 [10] used the following scenarios. The interactivity decreases with increasing Task No.:

- Task 1: Take turns reading random numbers aloud as quickly as possible
- Task 2: Take turns verifying random numbers as quickly as possible
- Task 3: Words with missing letters are completed with letters supplied by the other talker
- Task 4: Take turn verifying city names as quickly as possible
- Task 5: Determine the shape of a figure described verbally
- Task 6: Free conversation

Kitawaki used trained participants who were told to focus on the delay when judging the conversational quality.

The quality scale ranged from 0 (bad quality) to 4 (excellent quality) which is different to the quality scale as defined in [1]. In order to be able to compare Kitawaki's results with the results from more recent experiments his results were simply shifted by one MOS. The language used was Japanese.

Hammer [2, 3] used a similar methodology to Kitawaki's. The test scenarios are sorted in the order of decreasing interactivity:

– Task iSCT – Interactive Short Conversation Test
– Task RNV – Random Number Verification
– Task aSCT – Asymmetric SCT
– Task FC – Free Conversation

Hammer used untrained test persons, and the examined factor (delay) was not disclosed to them. The language used was German. The SAR parameter was evaluated in range of 40 (for RNV) to 15 (FC).

Kitawakis's and Hammer's results (see Fig. 1) indicate that Kitawaki's subjects seem to be more sensitive to delay artifacts than in Hammer's experiments. This chapter tries to find an explanation for these differences with the help of a new subjective conversational experiment. The new experiment will be discussed in more detail later in the Chapters 3 and 4.

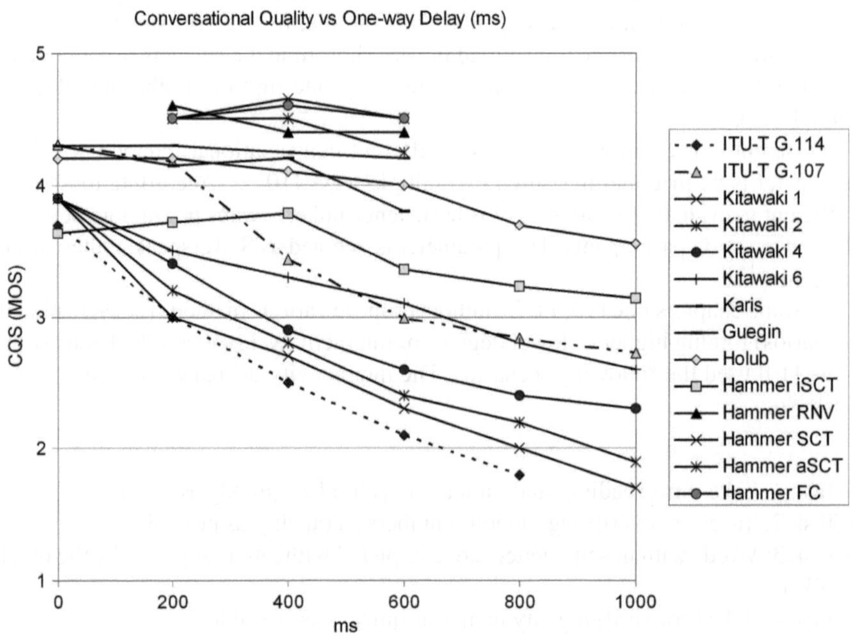

Fig. 1 The dependency of subjective conversational quality on transmission delay in echo-free situations – experimental results across different laboratories and different conversational tasks

2 Performed Tests

Based on the differences between the results from Kitawaki and Hammer new subjective tests were designed and performed. The purpose of these tests was to better understand the reasons of those differences. Twenty-two untrained test persons and four experienced interlocutors participated in the tests. The task of the interlocutor was to keep the conversation alive. The test persons were of age 19–38 and males and females were represented equally.

Four conversational scenarios forcing a rather stable degree of interactivity (SAR app. 15–20) were used:

- Ordering pizza from pizzeria;
- Ordering holiday trip from travel agency;
- Ordering air tickets from airlines;
- Ordering ticket to cinema.

The language used was Czech.

Each conversation lasted typically 2–3 min. In total, more than 400 conversations were performed and recorded; with one-way delays of 0, 150, 300, 600, and 900 ms, randomly distributed among conversations. After each conversation, both test person and interlocutor were giving the following four opinion scores in the given order:

- Listening quality;
- Talking quality;
- Interactive quality;
- Conversational quality (reported further in the article).

The test person was seated in an anechoic chamber with a reverberation time of less than 200 ms and a background noise of less than 30 dB SPL (A). The interlocutor was seated in a regular office environment with basic acoustic measures like sound absorbing lining and furniture. The background noise there was under 30 dB SPL (A) as well.

For about 50% of the test conversations, the background Hoth noise was artificially generated at both sides at a level of 56 dB SPL (A). PSTN and VoIP narrowband connections were simulated by a network simulator. Further details about the test design and conditions are available in [9].

3 Test Results

The results were averaged per delay condition. Conditions with background noise did not exhibit any important differences compared to noise-free conditions in echo-free scenarios. The results are summarized in Fig. 1.

The test conditions where the talker echo was presented are shown in Chapter 6. There are some arising questions in these cases which are discussed later as well.

4 Discussion

The results (Fig. 1) show a strong influence of different transmission delays (one-way) on the MOS-CQS values: The quality decreases with increasing delay. This behavior corresponds well to Hammer's [2, 3] and Guegin's [1] results. The results are considered as similar because the slopes of the graphs match strongly.

The quality difference in MOS-CQS between 0 and 400 ms delay in Kitawaki's experiment [10], as predicted by algorithmic models [5] or as shown in [4] is about 1 MOS point. However, in the new experiment and the experiments [1–3, 8, 9, 11] the influence of the transmission delay does not exceed 0.2 of on the MOS scale.

A possible reason for those differences may be due to the highly interactive scenarios used in Kitawaki's experiments but may be also influenced by the fact that trained test persons (explicitly focused to delay) were used.

Another explanation could be that telecommunication users became in general more tolerant to delay effects due to more and more deployed packet-based services like digital mobile networks or VoIP-based networks. However, the results of Karis [8] do not support this hypothesis since his experiments were conducted also in the 1991 s as Kitawaki's work.

5 Test Conditions with Talker Echo

Talker Echo (TE) is considered as important impairment of contemporary telecommunication networks. In the past, several subjective tests examining echo perception

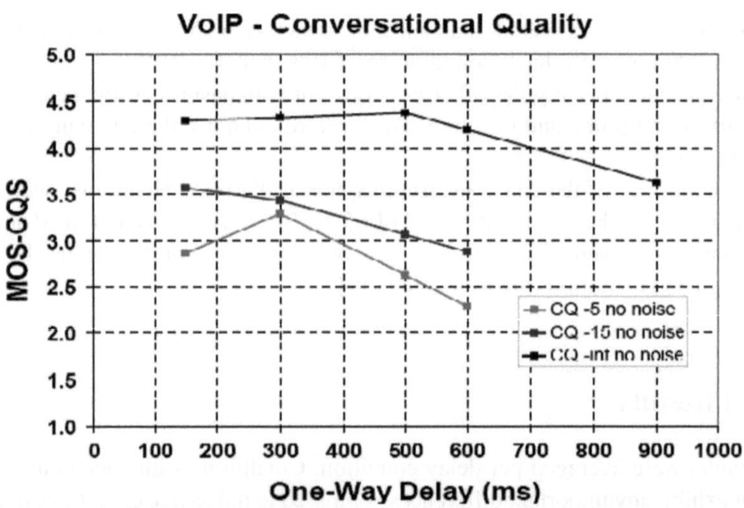

Fig. 2 Subjective test results in Czech experiment. The results for –5 dB ERL are non-monotonic. Data for delay of 0 ms are not available, however, it is highly presumable all three curves converge to single value of approximately 4.5 MOS

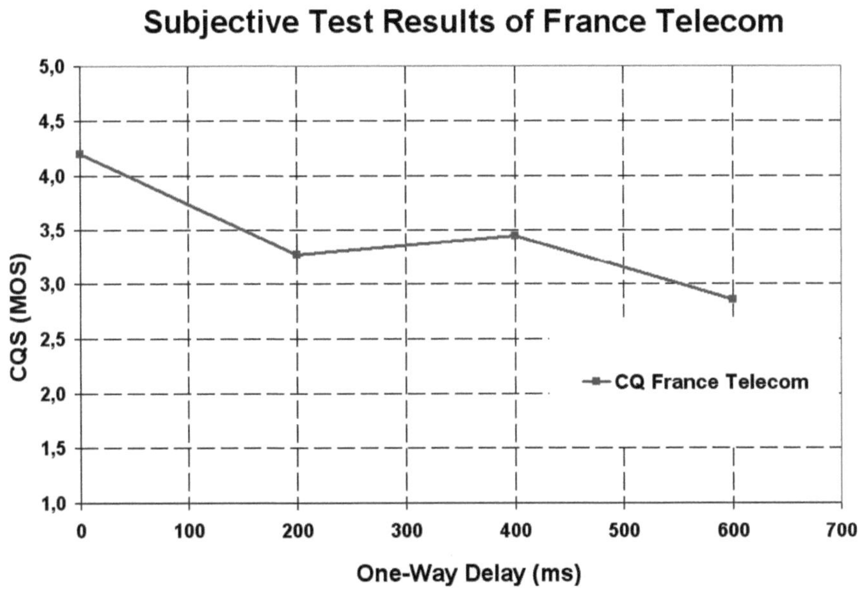

Fig. 3 Subjective test results of France Telecom experiment [1]. The results are again (like ERL–5 db in Fig. 2) non-monotonic

have been carried by different laboratories for different purposes, e.g., [1] and [9]. The results in case of echoed conversation cases show quite non-monotonic subjective assessment – in the range of cca 200 ms one-way delay (corresponding to Echo Delay ED of about 400 ms), the subjective perception is quite worse than in the area of 300–400 ms of one-way delay (600–800 ms echo delay). As both mentioned tests had quite high result variance (not shown in the picture), this non-monotonicity can be easily explained by measurement error caused by low number of tests participants. And – this used to be done so far.

From observation of Figs. 2 and 3, obvious question arises: Is the shown non-monotonicity of CQS versus delay in case of rather strong echo random effect that occurred in both laboratories independently in the same delay position?

More subjective test results should be analyzed to study this topic further. The problem is that raw measurement data are usually not published and only final regression (that does not contain possible original non-monotonicity any more) is available. We assume this is also a way how [4] has been derived.

6 Conclusion

Our experiments show that contemporary conversation participants show low sensitivity to transmission delay up to 500 ms (one-way) in echo-free connections. A comparison with the results of other laboratories indicated also that for common conversation scenarios this sensitivity does not increase significantly for any interactivity level evoked during the test.

An original question about non-monotonicity of CQS versus delay results for highly echoed communications is raised. Its answer requires deeper analysis of already available subjective tests and also carrying new subjective tests focused on this aspect. In case such subjective non-monotonicity is confirmed, it should be reflected properly in future versions of objective models and algorithms.

Acknowledgments This work has been supported by the Czech ministry of Education: MSM 6840770014 "Research in the Area of the Prospective Information and Navigation Technologies."

References

1. Guegin M, Gautier-Turbin V, Gros L, Barriac V, Le Bouquin-JEannes R, Faucon G (2005) Study of the Relationship between Conversational Quality, and Talking, Listening and Interaction Qualities: towards an Objective Model of the Conversational Quality, Measurement of Speech and Audio Quality in Networks MESAQIN 2005, Prague
2. Hammer F (2006) Quality Aspects of Packet-Based Interactive Speech Communication, Dissertation Work, Technical University Graz
3. Hammer F, Reichl P, Raake A (2004) Elements of Interactivity in Telephone Conversations, 8th International Conference on Spoken Language Processing (ICSLP/INTERSPEECH 2004), Jeju Island, Korea
4. ITU-T Rec. G.114 (1993/2003) One-way Transmission Time, International Telecommunication Union, Geneva
5. ITU-T Rec. G.107 (1996/2005) The E-model, a Computational Model for Use in Transmission Planning, International Telecommunication Union, Geneva
6. ITU-T Rec. P. 800 (1996) Methods for Subjective Determination of Transmission Quality, International Telecommunication Union, Geneva
7. ITU-T Rec. P. 800.1 (2006) Mean Opinion Score (MOS) Terminology, International Telecommunication Union, Geneva
8. Karis D (1991) Evaluating Transmission Quality in Mobile Telecommunication Systems using Conversational Tests. In: Human Factors Society 35th Annual Meeting, Santa Monica, CA, 1, 217–221
9. Kastner M, Holub J, Tomiska O (2006) Subjective and Objective Conversational Quality Assessment, ETSI STQ Mobile #13, Prague
10. Kitawaki N, Itoh K (1991) Pure Delay Effect on Speech Quality in Telecommunications, IEEE J. Sel. Areas Comm., 9, 586–593
11. Moeller S, Jekosch U, Raake A (2000) New Models Predicting Conversational Effects of Telephone Transmission on Speech Communication Quality, Proc. Int. Conf. Spoken Language Processing (ICSLP2000), CHN-Beijing, Vol. II, pp. 190–193

Performance Evaluation of EVRC-Encoded Voice Traffic over CDMA EVDO Rev. A

Fulu Li, Ivan Vukovic, Igor Filipovich, Phil Fleming, Eric Chan, and Andrew Lippman

Abstract Although CDMA EVDO Rev. A provides significant improvements over EVDO Release 0 that make it well suited for VoIP-like applications [3], it remains unclear about its performance in terms of quantitative evaluations with different real implementation scenarios. To understand the system design tradeoffs, we first investigate the traffic characteristics of EVRC-encoded real voice traces and we find that the correlation structure of voice traffic exhibits strong long-range dependency. We further observe that the burst length distribution of voice traffic is heavy tailed. In the major part of our study, we examine the tradeoff between the initial playout delay and the voice quality for real voice traces over CDMA EVDO Rev. A. We also examine the tradeoff between voice quality and different RF conditions. Additionally, we address the performance aspects of voice traffic over CDMA EVDO Rev. A with adaptive frame bundling (AFB) techniques and the impact of different treatments of the eighth rate frames for EVRC-encoded voice traffic. Finally, we examine the impact of hard handoff and random frame errors on voice quality for EVRC-encoded voice traffic over CDMA EVDO Rev. A.

1 Introduction

One of the most important services of today's wireless communication networks is voice service [1], and voice service-based revenues are still the dominant income sources for cellular service providers. With the deployment of CDMA EVDO Rev. A, more efficient solutions are needed for VoIP (voice over IP) over CDMA EVDO Rev. A to provide graceful migration from CDMA2000 1 x [2] to CDMA EVDO Rev. A [3]. Although CDMA EVDO Rev. A provides significant improvements over EVDO Release 0 that make it well suited for VoIP-like applications [3], it remains unclear about its performance in terms of quantitative evaluations with different real implementation scenarios. In this chapter, we conduct extensive simulations

F. Li (✉)
The Media Laboratory, Massachusetts Institute of Technology, Cambridge MA, 02139, USA
e-mail: fulu@mit.edu

S. Powell, J.P. Shim (eds.), *Wireless Technology*, Lecture Notes in Electrical Engineering 44, DOI 10.1007/978-0-387-71787-6_7,

on performance evaluation of EVRC (enhanced variable rate codec)-encoded [7–9] voice traffic over CDMA EVDO Rev. A in a variety of circumstances. We hope that the findings in these empirical studies could shed some light on more efficient solutions for VoIP over CDMA EVDO Rev. A.

In CDMA EVDO Rev. A, the forward link of the system consists of a single data channel that is divided into 1.67 ms time slots [4]. It is a time-multiplexed channel. The scheduler attempts to exploit the temporal variations of the channel by scheduling transmissions to mobiles during which time periods that the mobiles experience strong signal levels, e.g., the multi-user diversity benefit [5].

The data rates supported on the forward link are 38.4, 76.8, 102.4, 153.6, 204.8, 307.2, 614.4, and 921.6 kbps and 1.2, 1.8, and 2.4 Mbps (totally 11 different data rates) [4]. It is a packet-based variable rate traffic channel. One of three modulation schemes QPSK, 8PSK, and 16QAM is used, depending on the data rate. Packet sizes range from 1 to 2 k bits for QPSK, 3 k bits for 8PSK to 4 k bits for 16QAM. A forward link packet may occupy from 1 up to 16 time slots, depending on the data rate. A scheduler at the access point (AP, base station) determines the order in which the access terminals (AT, mobiles) are served. We will have more discussions on the scheduling block in Section 2.2. Two pilot bursts are inserted into each time slot to aid in synchronization, signal-to-interference plus noise ratio (SINR) estimation and coherent demodulation.

In our experimental studies, we use OpNet simulator [6] to model the channel conditions of each VoIP user, where path loss, large-scale shadowing, and small-scale temporal fading are included in the SINR (signal-to-interference plus noise ratio) calculation. For each mobile, a data rate (DRC – data rate control) is determined every 1.67 ms based on the SINR of the current channel condition. A scheduler uses these DRCs to determine the order and the data rate in which to transmit data to users on the forward link. We will have a brief discussion on the scheduling algorithms in Section 2.2.

In one of our experimental scenarios, we consider the case where each mobile can take advantage of receiver antenna diversity versus the case where mobiles are not equipped with dual receiver antennas. The dual antenna systems use dual receivers to exploit the antenna diversity gains and the pilot-weighted combiner is used in our implementation.

The rest of this chapter is organized as follows: The main thrust of the chapter is presented in Section 2. We show the characteristics of EVRC-encoded voice traffic in Section 2.1. The performance evaluation methodology is presented in Section 2.2. We present experimental results on VoIP over CDMA EVDO Rev. A in a variety of circumstances in Section 2.3. We discuss future trends in this area in Section 3. The conclusions are given in Section 4.

2 Main Thrust of the Chapter

We first investigate the characteristics of EVRC-encoded real voice traffic and then conduct a comprehensive empirical study on EVRC-encoded VoIP traffic over CDMA EVDO Rev. A in a variety of circumstances.

2.1 EVRC-Encoded Voice Traffic

In this section, we investigate the characteristics of EVRC-encoded voice traffic. EVRC stands for enhanced variable rate codec, which is widely used for CDMA cellular systems [7, 8, 9]. A Codec is essentially a data compression algorithm and the idea is to stuff as much information into as small a stream of data as possible [8]. It compresses each 20 miliseconds (ms) of 8000 Hz, 16-bit sampled speech input into output frames of full rate (171 bits), half rate (80 bits) or the eighth rate (16 bits). A frame sequence of EVRC-encoded real voice trace is shown in Fig. 3. EVRC packet consists of 12 binary words, each word 16 bits wide. The first word in the packet describes the type (4 = full rate, 3 = half rate, 1 = eighth rate, 0xE = erasure) and the other 11 words contain the actual packet data, padded with zeros if necessary to complete the octets (no zero padding for half rate and eighth rate frames).

We define the correlation co-efficiency among the frame sequence of EVRC-encoded voice traffic as

$$\rho_k = \frac{1}{N-k}\sum_{i=1}^{N-k}\frac{(x_i-\bar{x})(x_{(i+k)}-\bar{x})}{\delta_x^2} \tag{1}$$

where N is the number of frames in the voice trace, x_i is the number of bits in the *ith* frame, and k is the lag index.

In Fig. 1, we show the correlation structure of six reliable conversational voice traces (one with 48,068 frames and the other five traces, each of which has 60,000 frames) and a theoretical 2-state Markov model. The Y-axis is the lag index (the index lag between two packets) and the x-axis is the value of correlation co-efficiency. As we can see that even the lag index goes to 200, the correlation co-efficiency is still around 0.1. Clearly, the correlation structure of voice traffic exhibits strong long-range dependency. As we can see that the correlation structure of the 2-state Markov model at the lower left corner could not accurately capture the dynamics of the voice traffic.

We next examine the burst length distribution of the EVRC-encoded voice traffic, which is shown in Fig. 2. The x-axis is the value of the burst length and the y-axis is the PDF (probability density function). The x-axis and y-axis are both plotted in log scale. In Fig. 2, we also plot the curves of a corresponding power law distribution and a corresponding geometric distribution based on curve fitting techniques. As we can see that the burst length of EVRC-encoded voice traffic exhibits *heavy-tailed* distribution, meaning a high-frequency population of burst lengths is followed by a low-frequency population that gradually *tails off asymptotically*. Therefore, efficient solutions for VoIP over CDMA EVDO lies in how to take advantage of the self-similarity nature of the voice traffic to avoid consecutive packet losses and how to adaptively bundle multiple frames together to improve the system utilization as voice frame sizes are typically small.

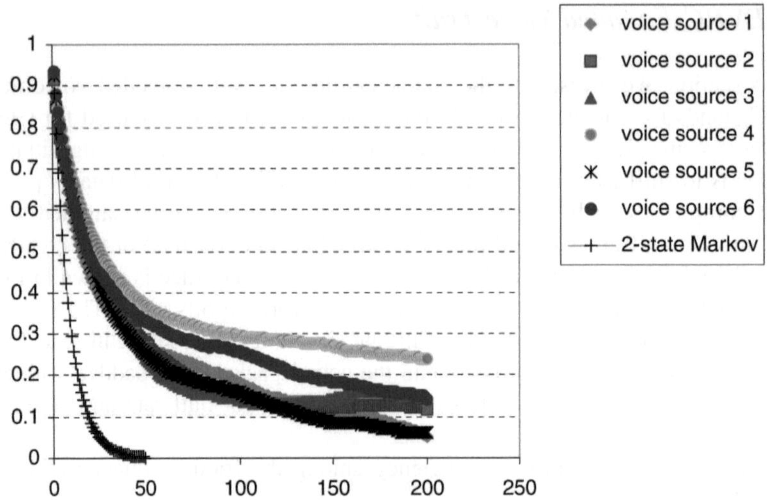

Fig. 1 The correlation co-efficiency of six EVRC-encoded conversational voice traces and the correlation co-efficiency of a trace generated by a 2-state Markov model

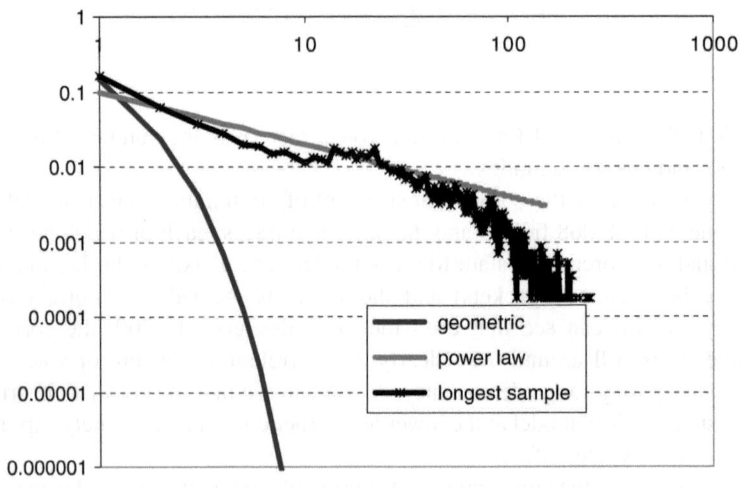

Fig. 2 The burst length distribution of EVRC-encoded voice traffic in a log–log scale versus that of geometric and power law distributions

The recent work in [1] on fractal analysis and modeling of VoIP traffic by Dang et al. also suggests that both VoIP call holding time and on/off period follow heavy-tailed distribution.

An illustration of the on/off pattern of an EVRC-encoded real voice trace is given in Fig. 3. In Fig. 3 the x-axis is the frame sequence and the y-axis is the frame size in bytes (full rate frame – 22 bytes, half rate frame – 10 bytes, eighth rate

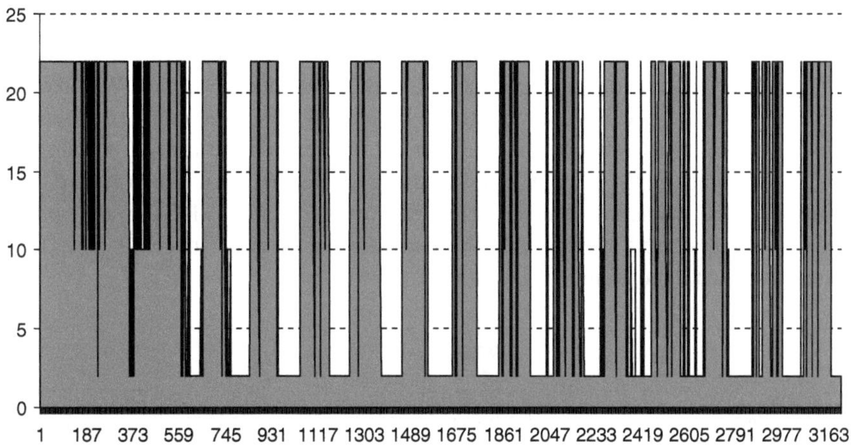

Fig. 3 An illustration of the on/off pattern of an EVRC-encoded real voice trace

frame – 2 bytes). The total number of frames in this example is 3,200. In this example, 60% of the whole period is busy and 40% of the whole period is idle.

2.2 The Evaluation Methodology

There are inherent differences between data traffic and voice traffic. First, data traffic may not exhibit the burst and silence nature of voice traffic (see Fig. 3 for an illustration). Second, voice traffic often has stringent delay constraints. We first examine the tradeoff between initial playout delay and voice quality. We use PESQ (perceptual evaluation of speech quality) software to get the MOS (mean opinion score) value. Voice quality was traditionally reported with a MOS value on a scale from 1 to 5 where 1 is the lowest (bad) and 5 is the highest (excellent). We illustrate the timing of frame generation, reception, and the initial playout delay concepts in the Table 1.

As shown in the following table, each frame is generated every 20 ms as the output of an EVRC encoder and the initial playout delay is set as 200 ms in this example. As we can see that the third frame arrives late for the playout and an

Table 1 An illustration of the timing of traffic generation, frame reception, and initial playout delay (200 ms playout delay in this example)

Frame seq	Generation (in msec)	Reception (in msec)	Playout (in msec)
1	0	180	200
2	20	205	220
3	40	245	240
4	60	255	260

erasure frame is submitted to the EVRC decoder instead, which could affect the quality of the voice perceived by the receiver.

The operation flow of the experimental studies is shown in Fig. 4. We take the original speech data, which are fed into the EVRC encoder. The output of the frame sequence from EVRC encoder is the input data for each VoIP user in our simulation. We simulate CDMA EVDO Rev. A using OpNet [6] simulator software. The distorted EVRC frame file received at the receiver is fed into EVRC decoder, whose output is the distorted speech data. We compare the original speech data and the distorted speech data using PESQ software.

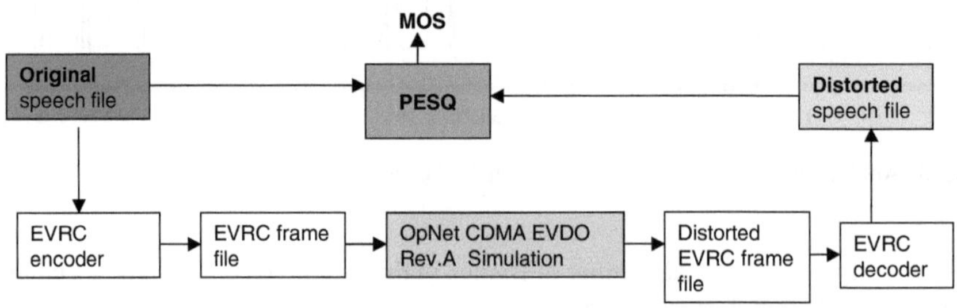

Fig. 4 The operation flow of the experimental studies

In our experiments, only some of the eighth rate frames are sent over the air, which is a common practice as sending all the eighth rate frames is not necessary in terms of voice quality appreciation by the end users. We use the following formula to measure the packet loss rate:

$$P_{loss} = \frac{the \quad number \quad of \quad Erasure \quad frames}{total \quad number \quad of \quad transmitted \quad frames} \tag{2}$$

There are two possible reasons for an erasure frame: one is that the original packet was lost in the air and the other is that the original packet arrives late (after its corresponding playout time).

We implemented two types of schedulers: one is the pure proportional-fairness (PF) scheduler without QoS consideration and the other is weighted proportional-fairness (WPF) scheduler with delay multiplier, e.g., with QoS guarantees.

In proportional-fairness scheduling (PF) algorithm [4], the user with the highest ratio of $\frac{R_k(t)}{T_k(t)}$ out of all Nusers will receive transmission at each decision time. $R_k(t)$ is the current request rate at time slot t and $T_k(t)$is user k's average throughput in a past window. Ties are broken randomly. Any user with no data to send is ignored during the scheduling process. $R_k(t)$ is determined by the channel conditions $C_k(t)$. As stated in [4], $T_k(t)$ can be updated by an exponential filter as follows:

$$T_k(t+1) = (1 - \frac{1}{W}) \times R_k(t) + \frac{1}{W} \times r_k(t) \tag{3}$$

where W is the number of slots, each of which is 1.67 ms long, and $r_k(t)$ is the current transmission rate that user k receives at time slot t.

In weighted proportional-fairness (WPF) with delay multiplier scheduling algorithm, the BTS (base transceiver station) scheduler sort metric is defined as follows:

$$S_i(t) = W_i R_i^{\beta}(t) \times \left(\frac{1}{W_i^{\gamma}(t)} \oplus Kf(t, t_{HOL}^{i}) \right) \tag{4}$$

where W_i is class-based weight; $R_i(t)$ is the rate determined by user i's DRC (data rate control) feedback, e.g., channel condition; $W_i(t+1) = W_i(t) + (1-\alpha)(D_i(t) - W_i(t))$ is user i s average throughput and $D_i(t)$ is the effective RF rate in bits/s that user i is having at time t; $\oplus$ can be addition or multiplication; K is scaling coefficient used to balance WPF (weighted proportional fairness) and delay metric terms; $f(t, t_{HOL}^{i})$ is an increasing function of time and it takes arrival time of the packet at the head of user i's BTS queue as a parameter; α, β, γ are configurable parameters.

In WPF scheduling algorithm, users with the highest metric of $S_i(t)$ are picked for transmission and in case of multi-user transmissions users are picked in sequential order.

2.3 Performance Evaluation

In the first set of experiments, we conduct experiments in the following three major scenarios: (1) the cases with or without receiver antenna diversity, (2) the cases with different mobile locations and different congestion level, and (3) the cases with or without QoS, e.g., pure proportional-fairness (PF) scheduling or the weighted proportional-fairness (WPF) scheduling with delay multiplier.

The first set of performance evaluation results of EVRC-encoded voice traffic over CDMA EVDO Rev. A are shown in Figs. 5, 6, 7 and 8. We use weighted proportional-fairness (WPF) scheduling with delay multiplier in the experiments shown in Figs. 5 and 6. As we can see from Figs. 5 and 6 that with receiver antenna diversity (Fig. 6) and the same number VoIP users per sector, e.g., 30 VoIP users per sector in this example, the initial playout delay can be reduced by about 20–30 ms with virtually the same voice quality compared with the case without receiver antenna diversity (Fig. 5).

Comparing Fig. 6 with Fig. 8, we can also observe that with QoS consideration (Fig. 6), e.g., weighted proportional-fairness (WPF) scheduling with delay multiplier, the system can support 20 more VoIP user per sector with virtually the same voice quality compared with the case without QoS support (Fig. 8) in this example.

Notably, in all the circumstances with the increase of the initial playout delay, the packet loss rate goes down. This is because of the fact that a larger initial playout delay allows more voice frames to arrive in time. It is also true in all the scenarios that the closer is the mobile from the cell tower, the better are the channel conditions, the less is the packet loss rate, and the better is the voice quality.

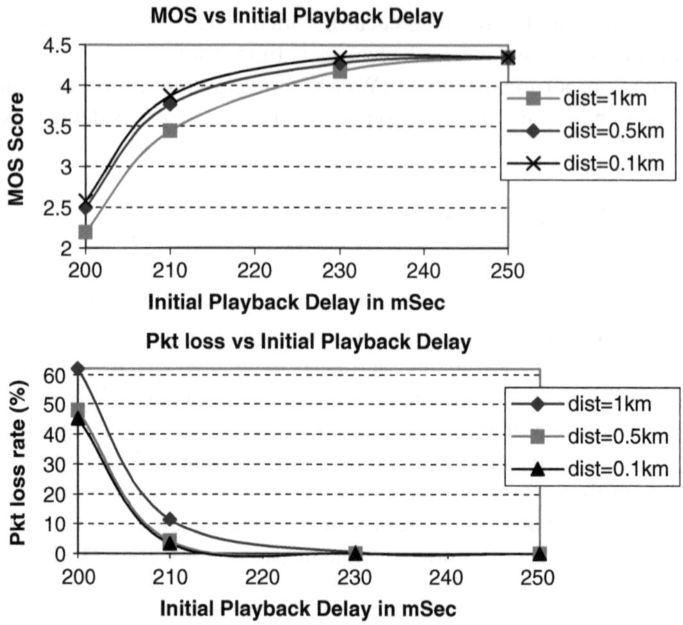

Fig. 5 VoIP over CDMA EVDO Rev. A (30 VoIP users per sector, without receiver antenna diversity, with QoS)

In Fig. 7, we show the performance comparison in terms of end-to-end delay and packet arrivals with different number of VoIP users per sector. The *x*-axis is the time sequence of the packet arrivals in minutes and the *y*-axis is the end-to-end delay in seconds in both of the upper and of the lower parts of the charts. The blue lines are the delay deadline, which is set as 0.18 s. The red dots are the packet arrivals with their *y*-coordinate values indicating their respective end-to-end delays. The upper part is the case with 3 VoIP users per sector and the lower part is the scenario with 135 VoIP users per sector in the same settings. Clearly, with fewer VoIP users per sector (the upper part in Fig. 7), more packets can arrive in time even with stringent delay deadline of 0.18 s. With more VoIP users per sector (the lower part in Fig. 7), fewer packets can arrive in time for playout (in this example, the majority of the packets cannot arrive in time with 135 VoIP users per sector). Clearly, more VoIP calls can be carried if the end-to-end delay target is longer. Even if not all the traffic is VoIP traffic the longer the end-to-end delay target the more other type of traffic (video, web, ftp) can be carried for the same QoS levels.

In the second part of our experiments, we evaluate the performance aspects of EVRC-encoded voice over CDMA EVDO Rev. A with adaptive frame bundling (AFB) techniques. In Fig. 9, the *x*-axis is the packet loss burst index and the *y*-axis is the packet loss burst length in terms of the number of frames. As we can see from Fig. 9 and 10 that packet loss occurs in burst and packet loss burst length exhibits heavy-tailed distribution. This may be due to the proportional-fairness

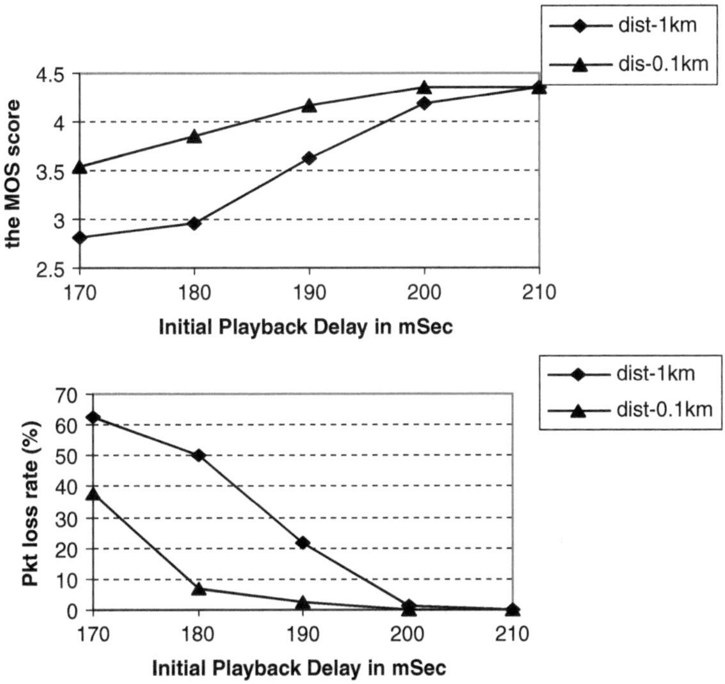

Fig. 6 VoIP over CDMA EVDO Rev. A (with 30 VoIP users per sector, with receiver antenna diversity, with QoS)

Fig. 7 Performance comparison in terms of end-to-end delay and packet arrivals with different number of VoIP users per sector

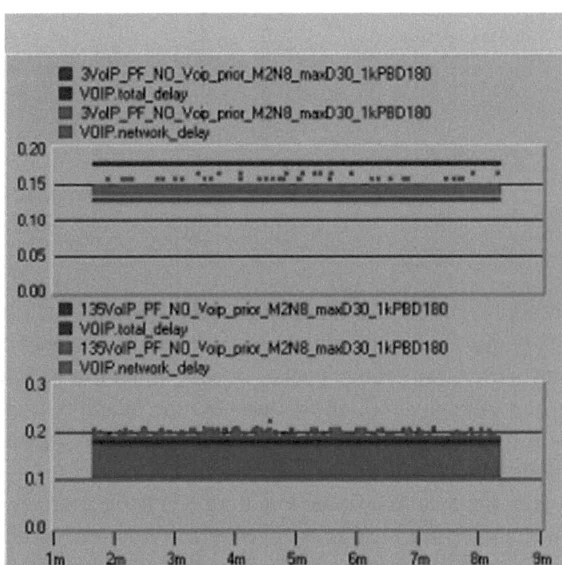

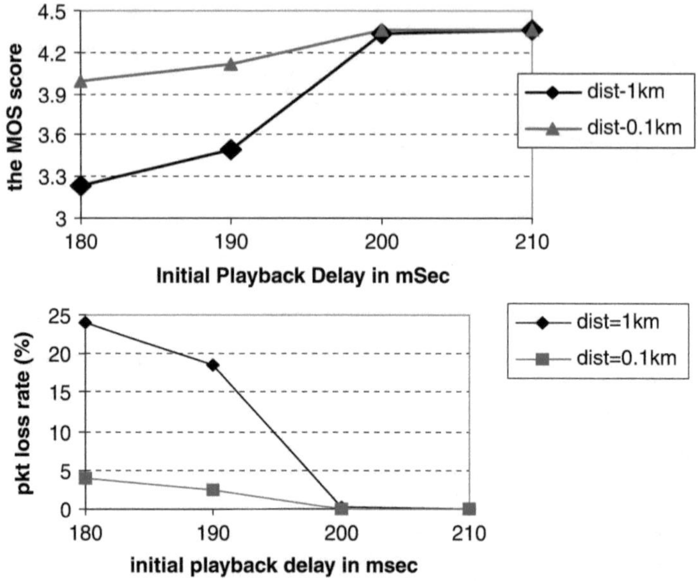

Fig. 8 VoIP over CDMA EVDO Rev. A (with 10 VoIP users per sector, with receiver antenna diversity, without QoS)

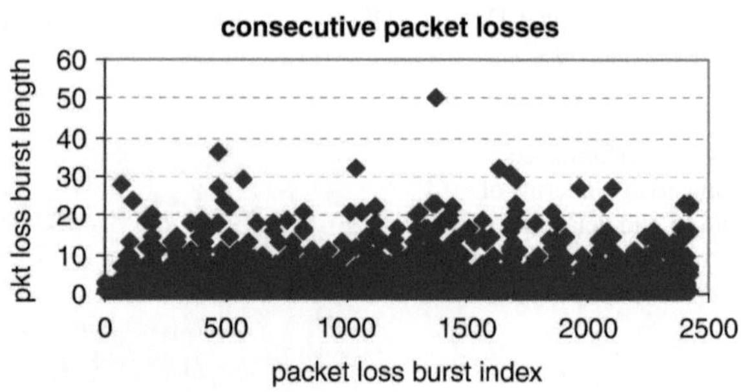

Fig. 9 An illustration of the consecutive packet losses for voice over CDMA EVDO Rev. A (for a specific VoIP user in our experiment)

nature [4] of CDMA EVDO to exploit the channel variations for multi-user diversity gains. However, some users with bad RF channel conditions, e.g., lower SINR, may suffer consecutive packet losses, which is a bad thing to avoid for VoIP users. In the following, we show that how the adaptive frame bundling techniques may help to alleviate consecutive packet losses for VoIP users with poor RF channel conditions.

For the adaptive frame bundling (AFB) techniques, in essence we want to wait and bundle multiple frames together for VoIP users with good RF channel conditions and exclude them from the consideration of current scheduling decision and push them into the next scheduling decision time slot with a bundled frame. This way it saves transmitting slots for other VoIP users with poor RF channel conditions

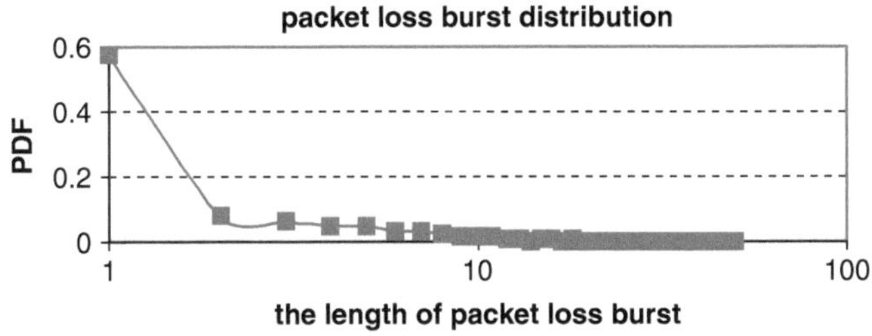

Fig. 10 An illustration of packet loss burst length distribution (the x-axis is the length of packet loss burst and the y-axis is the corresponding probability density function [PDF])

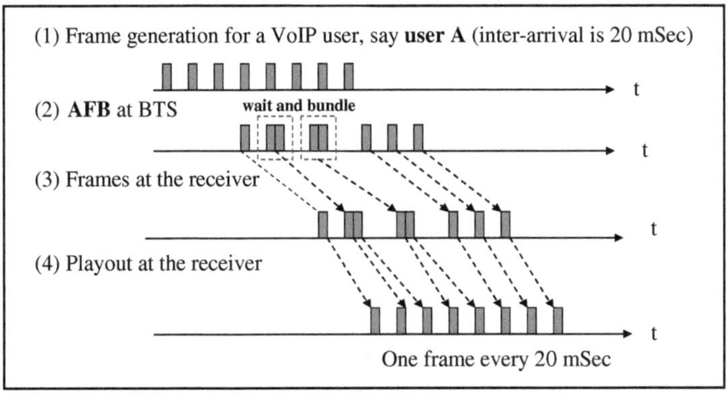

Fig. 11 An illustration of adaptive frame bundling (AFB) technique

which may be experiencing consecutive packet losses. Of course, the frame bundling operation has to be adaptive to RF channel conditions, head-of-line (HOL) frame delay constraints, buffer availability, and overall system traffic load.

An illustration of the AFB technique is given in Fig. 11. The voice frame is generated at one frame per 20 m at the source. At the BTS, it adaptively waits and bundles multiple frames together (two frames in this example). When the voice data are played out at the receiver, the bundled frames are separated just in time for playout.

An illustration of the benefits of adaptive frame bundling techniques to alleviate packet losses for VoIP users is given in Fig. 12. In Fig. 12, the y-axis is the percent of the users that have less than 2% frame losses and the x-axis is the end-to-end delay in seconds. Clearly, those with bad SINR have significant gains to avoid consecutive packet losses with the adaptive frame bundling technique (the red one). The idea is to flatten the distribution among all users with regard to frame losses. In this example, we can see that when the end-to-end delay is set in the range of 0.25–0.265 s, which is typically the case for current CDMA EVDO Rev. A systems, with adaptive frame bundling more VoIP users have less than 2% frame losses compared with approaches without frame bundling. The benefits of adaptive frame bundling techniques could be further improved with appropriate parameter tuning.

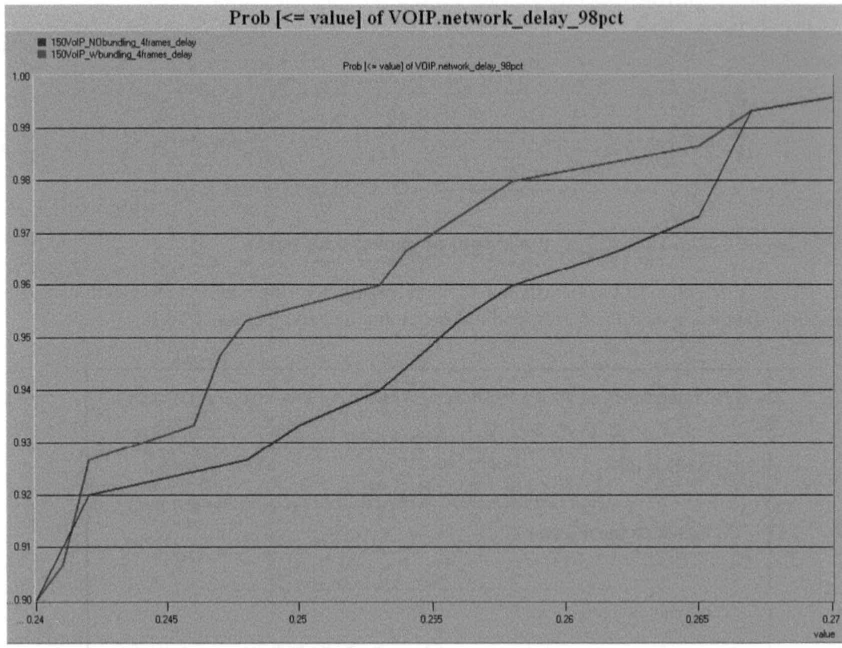

Fig. 12 Performance comparison for approaches with or without adaptive frame bundling (AFB) techniques in terms of packet loss

In the third part of our empirical study, we evaluate the impact of different treatments of the eighth rate frames for EVRC-encoded voice traffic over CDMA EVDO Rev. A and the impact of random frame errors as well as the impact of hard handoff (as proposed for 4 G technology) on voice quality for EVRC-encoded voice traffic over CDMA EVDO Rev. A.

In Fig. 13, we show the MOS scores versus different packet loss rates with different treatments of the eighth rate frames for EVRC-encoded voice traffic over CDMA EVDO Rev. A in the same settings. In this figure, the *x*-axis is the packet loss rate in percentage and the *y*-axis is the MOS scores of the received voice traffic at the mobile users-end. The bottom line (the red line) is the case without the eighth rate frames (indicated as "no ER" in the legends). The top line (the blue line) is the scenario with all the eighth rate frames (indicated as "w/ER" in the legends). The middle line (the orange line) is the case with partial eighth rate frames (indicated as "periodic ER" in the legends), where in silence periods one eighth rate frame is sent for every eight inter-frame-arrival periods, e.g., every 160 msec. Clearly, partially reducing the eighth rate frames leads to slightly lower MOS scores. Completely eliminating the eighth rate frames could result in a MOS score drop of 0.5 on an average. Therefore, for the eighth-rate-frame-suppressed speech its voice quality is worse than that of the non-eighth-rate-frame-suppressed speech.

On the other hand, reducing or eliminating the eighth rate frames for EVRC-encoded voice traffic could greatly increase the capacity of the system. This is due

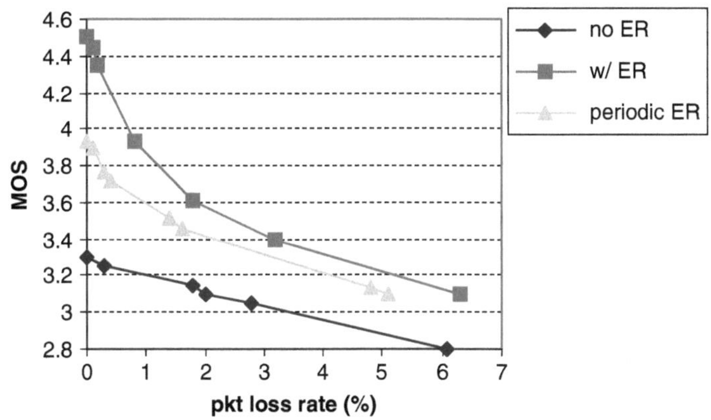

Fig. 13 MOS scores versus packet loss rate with different treatments of the eighth rate frames for EVRC-encoded voice traces

to the fact that on an average nearly 60% of the EVRC-encoded voice traffic consists of eighth rate frames because of the on/off pattern of the voice traffic. The reduction or elimination of the eighth rate frames will significantly lower the effective rate of each VoIP user's voice traffic, which will lead to a larger cell capacity in the sense that each cell can accommodate more VoIP users.

Lastly, we evaluate the impact of hard handoff and random frame errors on the voice quality for EVRC-encoded voice traffic over CDMA EVDO Rev. A.

In Table 2, we show VoIP MOS scores with or without handoff and a combination of different random frame error rates (FER) in the same settings. The right column depicts the PESQ MOS score ranges that are obtained during our empirical studies. The central column shows the average PESQ MOS score values. Notably, the impact

Table 2 VoIP MOS scores with or without handoff and a combination of different random frame error rates (FER)

	Average PESQ MOS	PESQ MOS range
0% Random FER, no handoff	4.5 (by definition)	4.5 (by definition)
0% Random FER, with handoff (one consecutive frame outage)	4.31	3.66–4.5
0% Random FER, with handoff (two consecutive frames outage)	4.26	3.44–4.5
0% Random FER, with handoff (three consecutive frames outage)	4.19	3.21–4.5
1% Random FER	3.9	3.9
2% Random FER	3.89	3.13–4.48
2% Random FER, with hard handoff (three consecutive frames outage)	3.76	3.03–4.48
3% Random FER	3.4	2.89–4.48

of up to three consecutive frame errors does not appear to be as significant as 1% random frame errors. Hard handoff has been proposed for 4G technology and we observe that with EVRC the impact of hard handoff is only about 0.1 in PESQ MOS scores. Thus, the degradation of voice quality caused by hard handoff is not as significant as random errors due to the design of EVRC itself.

In summary, we conducted comprehensive empirical studies for EVRC-encoded voice traffic over CDMA EVDO Rev. A in a variety of circumstances to provide quantitative evaluations to better understand the system design tradeoffs.

3 Future Trends

We believe that the use of dual receiver antennas at end-user mobiles will become commonplace due to the benefits of the receiver antenna diversity that we have seen in this empirical study. It is also possible for the use of self-adjusted directional antennas for better signal strength and more efficient power usage at the end-user mobiles.

Moreover, it is beneficial for the improvement of the overall QoS among all the VoIP users with the employment of adaptive frame bundling (AFB) techniques at the BTS (base transceiver station) scheduler as discussed in this study.

We also foresee that smarter scheduling algorithm at the BTS is needed to better balance the QoS requirements among all VoIP users as well as to better exploit the multi-user diversity based on dynamic channel conditions and the cooperation among different end-user mobiles.

Further, smarter transceiver coupled with multi-connectivity capability at the end-user mobiles to dynamically switch its associated network for better signal coverage and/or cost efficiency is a better way to go. For example, a WiFi-enabled cell phone may be able to dynamically switch from its associated cellular network to a WiFi hotspot to make a phone call based on its environment. It is also possible for cooperative relay among different end-user mobiles for better coverage and more efficient power usage.

Lastly, the use of more advanced codec [7, 8, 9] for the encoding/decoding of voice traffic and the adoption of VoIP packet header compression [10] can further improve the system capacity, which will lead to better services for VoIP users.

4 Conclusion

Although CDMA EVDO Rev. A provides significant improvements over EVDO Release 0 that make it well suited for VoIP-like applications [3], it remains unclear about its performance in terms of quantitative evaluations with different real implementation scenarios.

To understand the system design tradeoffs, we investigate the traffic characteristics of EVRC-encoded real voice traces and we conduct extensive experiments for performance evaluation of EVRC-encoded voice traffic over CDMA EVDO Rev. A in a variety of circumstances. We hope that the findings in these empirical studies could shed some light on more efficient solutions for VoIP over CDMA EVDO Rev. A. We will investigate the performance of voice traffic over CDMA EVDO Rev. A with the consideration of user mobility as our future directions.

Acknowledgment The authors would like to thank Jim Ashley for the voice traces, Mike Kirk and Raghu Hariharan for PESQ tool and some other tools to measure MOS score, Edgardo Cruz for EVRC software, Rangsan Leelahakriengkrai for various support during this project. The authors would also like to thank Mehmet Yavuz and Chong Lee at Qualcomm Inc. for valuable comments on the paper. Lastly, Fulu Li and Andrew Lippman would also like to thank the Digital Life consortium at MIT Media Lab for the support.

Reference

1. T. Dang, B. Sonkoly, S. Molnar, "Fractal Analysis and Modeling of VoIP traffic" in the proceeding of NETWORKS '2004.
2. VoIP over 1X EVDO at http://www.airvananet.com/files/VoIP_over_EVDO.pdf
3. M. Yavuz, S. Diaz, R. Kapoor, M. Grob, P. Black, Y. Tokgoz, C. Lott, "VOIP over CDMA2000 1xEV-DO Revision A", IEEE Communications Magazine, Feb. 2006.
4. A. Jalali, R. Padovani, R. Pankaj, "Data Throughput of CDMA-HDR: A High-Efficiency High-Data-Rate Personal Communication Wireless Systems", in the Proc. of IEEE VTC 2000 (Spring).
5. D. Tse, "Forward Link Multiuser Diversity Through Rate Adaptation and Scheduling", submitted for publication to IEEE JSAC.
6. http://www.opnet.com
7. Enhanced Variable Rate Codec (EVRC) at http://en.wikipedia.org/wiki/EVRC
8. EVRC: the Savior of CDMA at http://www.arcx.com/sites/EVRC.htm
9. M. McDonald, "EVRC: Best of Both Worlds", at http://telephonyonline.com/wireless/mag/wireless_evrc_best_worlds/
10. J. Ash, B. Goode, J. Hand, " Requirements for End-to-End VoIP Header Compression", at http://www.ietf.org/proceedings/03jul/slides/avt-12.pdf
11. P. Mehta, S. Udani, "Overview of Voice Over IP", Technical Report, Univ. of Pennsylvania, Feb. 2001.

Efficient Structures for PLL's Loop Filter Design in FPGAs in High-Datarate Wireless Receivers – Theory and Case Study

Yair Linn

Abstract In most contemporary phase lock loops (PLLs) used in high-datarate wireless receivers, some or all of the PLL's components are implemented digitally, in particular the PLL's loop filter. In this chapter we develop the theory behind new efficient structures for the implementation of loop filters within FPGAs (Field Programmable Gate Arrays) using fixed-point arithmetic. The theory is then investigated via a case study, in which we present FPGA hardware mapping results that show that employing the proposed method results in a decrease of more than 70% in the logic gate count needed as compared to the conventional implementation.

1 Introduction

Receivers in modern communications systems often contain several phase lock loops (PLLs). For example, in a coherent wireless communications system the receiver contains at least two PLLs, namely one that performs carrier synchronization and another that is tasked with symbol timing recovery.

In most modern systems, some or all of the PLL's components are implemented digitally, in particular the loop filter. In this chapter we develop efficient structures for the implementation of loop filters within FPGAs (field programmable gate arrays).

We start by deriving equations that mathematically describe the loop filter's characteristics as a function of the PLL's performance requirements. We then discuss some digital filter topologies that are suitable for efficiently implementing the loop filter in FPGAs using fixed-point arithmetic. For the case of high-speed communications (i.e., with datarates of at least 1 MegaSymbols/Second), it is found that the Direct-Form II topology can be exploited to yield an ultra-compact

Y. Linn (✉)
Universidad Pontificia Bolivariana, Bucaramanga, Colombia
e-mail: yairlinn@gmail.com

S. Powell, J.P. Shim (eds.), *Wireless Technology*, Lecture Notes in Electrical Engineering 44, DOI 10.1007/978-0-387-71787-6_8,

implementation. This is done by exploiting the fact that often the symbol rate of high-datarate systems is much higher (by several orders of magnitude) than the PLL's natural frequency. This attribute of the PLL allows us to substantially lower the clock rate at which the loop filter operates by using a decimator placed between the PLL's phase detector and the loop filter. The reduction in the loop filter's operating clock rate allows us to avoid direct implementation of the multiplication operations in the loop filter. Rather, each multiplication is implemented by a state machine that iteratively sums and shifts the partial products encountered during the multiplication process. An additional improvement (in terms of implementational efficiency) is achieved by modifying the Direct-Form II filter structure by introducing a pipeline register between certain filter elements.

We present FPGA hardware mapping results conducted using the Xilinx Virtex XCV600-4HQ240 chip, which show that employing the proposed method results in a decrease of more than 70% in the logic gate count needed as compared to the conventional implementation.

2 Receiver Model

The overwhelming majority of PLLs are of second-order [1–4]. This is because second-order PLLs are unconditionally stable [3 Section 2.4.2]. This is indeed the PLL type treated in this chapter. The linear-model transfer function of the second-order PLL in the Laplace domain is [5 Chapter 2]

$$H(s) \triangleq \frac{\theta_o(s)}{\theta_i(s)} = \frac{2\zeta\omega_n s + \omega_n^2}{s^2 + 2\zeta\omega_n s + \omega_n^2} \tag{1}$$

where θ_i is the input oscillator phase and θ_o is the PLL's local oscillator phase. From (1) it is clear that the second-order PLL is completely defined by its natural (radian) frequency ω_n and its damping factor ζ.

In Fig. 1 we see an example receiver structure to which the derivations of this chapter apply. It is stressed that though in Fig. 1 a hybrid carrier PLL is shown as an example, the derivations of this chapter are actually applicable to any hybrid or digital PLL for which the phase detector sample rate is much higher than the PLL's natural frequency. We write this condition as follows:

$$f_p \gg f_n \tag{2}$$

where $f_p = 1/T_p$ is the phase detector sample rate and $f_n(=\omega_n/2\pi)$ is the PLL's natural frequency. In carrier PLLs and in symbol timing recovery loops the PLL phase detector sample rate is the same order of magnitude as the symbol rate [1, 2], that is we have $1/T_p \sim 1/T$ where $1/T$ is the symbol rate and "$\sim$" denotes equal orders of magnitude. Conversely, the natural frequency of the PLL is the same order of magnitude as the significant bandwidth of the received

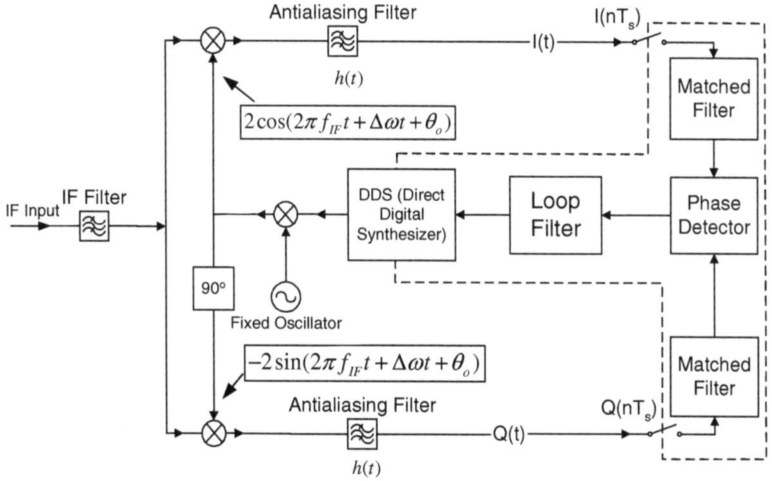

Fig. 1 General structure of a hybrid carrier PLL for digital wireless communications. The parts within the *dashed line* are implemented digitally, while the rest are analog components (the samplers and the DDS are mixed-signal components). $1/T_S$ is the sample rate. f_{IF} is the IF (intermediate frequency) and $\Delta\omega$ is the frequency difference between the local and the input oscillators ($\Delta\omega = 0$ when the PLL is locked)

carrier phase noise [6] (for carrier PLLs) or as the symbol clock phase noise (for symbol timing recovery PLLs). Carriers[1] that are used in coherent communications generally have phase noise whose content can be assumed non-negligible up to a distance of at most several kHz. To give an example, in the DVB-S2 standard [7], the specification is –68 dBc/Hz at 1 kHz offset from the carrier. Although f_n is depends upon the parameters of the particular communications system, for a great proportion of contemporary wireless communications systems that the *order of magnitude for f_n is in kHz*. Thus, for many practical cases we have that (2) holds.

3 The Equivalent Linearized Hybrid or Digital PLL Model

PLL analysis is customarily done using the equivalent linear baseband model, as shown in Fig. 2. This is a general model that is applicable to both hybrid and digital PLLs. In [8] a methodical approach for the design of hybrid PLLs was developed, and the loop filter design methodology outlined there is also applicable to the all-digital PLL. There, it was shown that due to (2) we can decimate the output of the phase detector before it enters the loop filter, and then implement the loop filter at a

[1] As for symbol timing clocks, these are usually derived from crystal oscillators, which also usually have negligible phase noise at a frequency offset of several kHz.

lower rate. That lower rate was called $f_u = 1/T_u$ Hz, and in [8] it is shown that we have

$$f_p \geq f_u \gg f_n \tag{3}$$

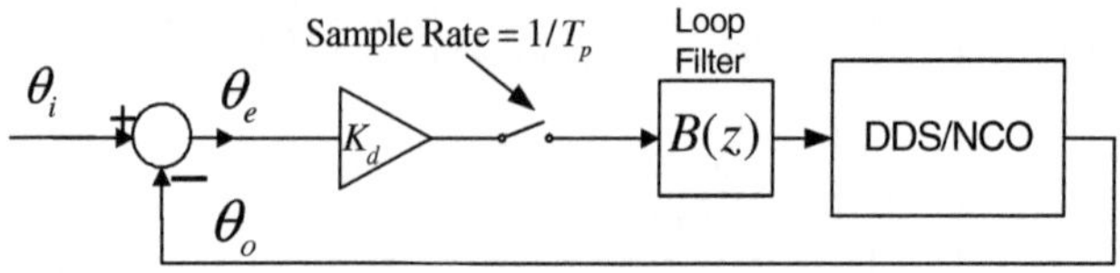

Fig. 2 PLL equivalent linearized baseband model. K_d is the phase detector gain. DDS = direct digital synthesizer (in hybrid PLLs). NCO = numerically controlled oscillator (for completely digital PLLs)

An equivalent baseband model of the PLL including this decimation is shown in Fig. 3. Note that in the actual implementation, the decimation filter and decimator will be inserted between the phase detector and the loop filter in Fig. 1. If we assume that the decimation process is ideal (a subject that is investigated in [8]), then for loop-filter analysis purposes we can simplify Figs. 3 and 4. Note that, unlike Fig. 2, in Fig. 4 the sample rate is $f_u = 1/T_u$.

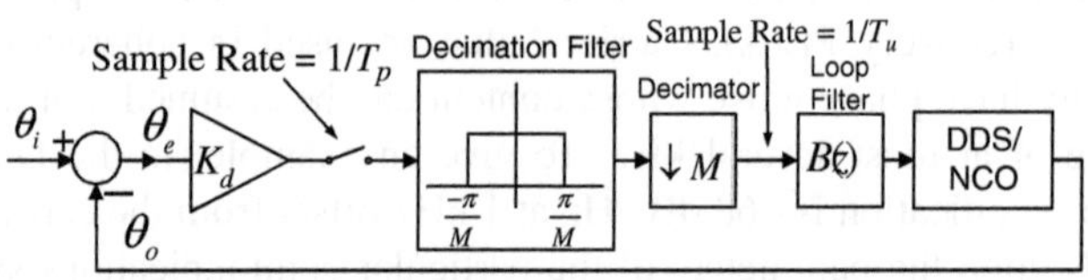

Fig. 3 –PLL equivalent baseband model with decimation before loop filter

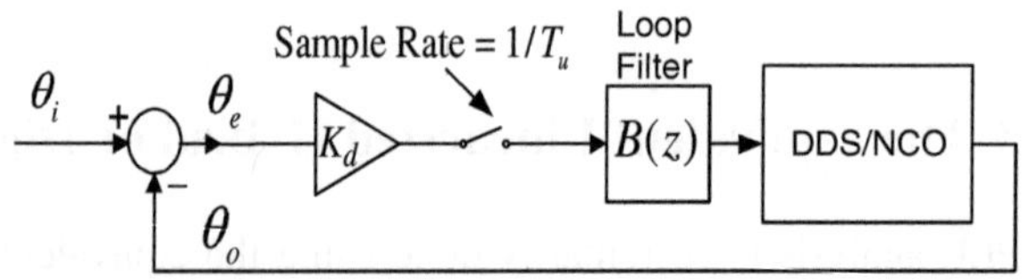

Fig. 4 Equivalent PLL model assuming that the decimation process is ideal. Note that the sample rate is $1/T_u$

4 The Analog PLL model – a Starting Point

As a starting point of our design, we shall shortly see that it is useful to first analyze an analog PLL [4 Chapter 2] with the same ω_n and ζ, and the design of the hybrid PLL of Fig. 2 will follow in short order.

4.1 The Analog PLL Loop Filter

We assume that the analog loop filter transfer function is

$$F(s) = K_a \frac{1 + s\tau_2}{1 + s\tau_1} \tag{4}$$

For discussion of other loop filters see [9, 10], and [11 Chapters 3–5]. Although the filter of (4) can be easily implemented as a resistor–capacitor network [4 Chapter 2], we will not discuss this implementation since we are interested in a *digital* implementation (see Figs. 1 and 2).

4.2 The VCO Transfer Function

In the Laplace domain, the VCO (voltage Controlled oscillator) transfer function is $V(s) \triangleq K_V/s$, where K_V is in units of rads/(sec · volt).

4.3 The Open and Closed-Loop Transfer Functions

The open-loop function is then (see Fig. 5)

$$G(s) = K_d F(s) V(s) = K_d F(s) K_V/s \tag{5}$$

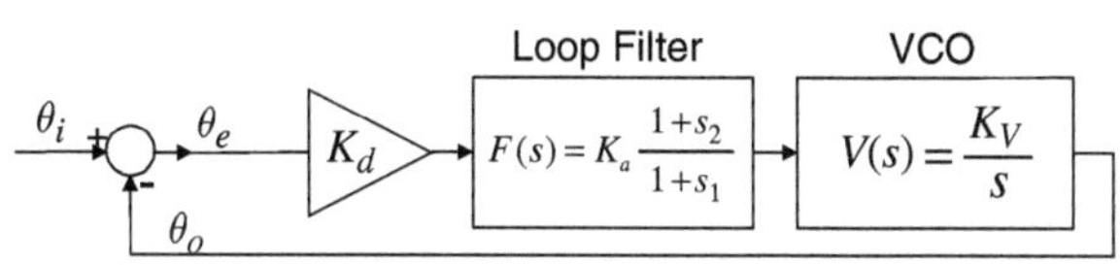

Fig. 5 Linearized analog PLL model

We define for convenience the *loop gain* as $K \triangleq K_d K_a K_V$. From (4) and (5) the closed-loop function of the PLL linear model is

$$H(s) = \frac{\theta_o(s)}{\theta_i(s)} = \frac{G(s)}{1 + G(s)} = \frac{K\frac{\tau_2}{\tau_1}s + \frac{K}{\tau_1}}{s^2 + \frac{1+K\tau_2}{\tau_1}s + \frac{K}{\tau_1}} \triangleq \frac{P(s)}{Q(s)} \tag{6}$$

This corresponds to the transfer function of a second-order system, whose denominator can be written as

$$Q(s) = s^2 + 2\zeta\omega_n s + \omega_n^2 \tag{7}$$

Comparing (6) and (7), we find that

$$\omega_n = \sqrt{K/\tau_1} \tag{8}$$

and using (6), (7), and (8)

$$\zeta = 0.5\sqrt{K/\tau_1} \cdot (\tau_2 + 1/K) \tag{9}$$

and substituting (8) and (9) into (6) we get

$$H(s) = \frac{(2\zeta - \omega_n/K)\omega_n s + \omega_n^2}{s^2 + 2\zeta\omega_n s + \omega_n^2} \tag{10}$$

We assume (as is usually the case) that the loop gain is high, i.e.,

$$K \gg \omega_n \tag{11}$$

There is no true loss of generality incurred by making this assumption, as virtually all PLLs are designed to have a high loop gain since this makes them relatively insensitive [12 Chapter 14] to variations in the gains K_d, K_a, and K_V.

Using (11), Eq. (10) can be reduced to (1), which is the model that we assumed.

The exact magnitude of the relationship described in (11) is usually determined by the allowable steady-state error [10, 11 Chapter 5]. To give an example, for a 2nd-order PLL with $F(s)$ of (4) the steady-state error to a frequency step of $\Delta\omega$ rads/sec will be $\Delta\omega/K$ rads [11 Chap. 5]. If we can assume that for the given communications system all frequency steps encountered follow $|\Delta\omega| \leq \omega_n$, then a constraint ensuring an acceptable upper bound on the steady-state error is, for example, $K = 100\omega_n$. Of course, different steady-state requirements will result in different instantiations for (11).

5 Digital Loop Filter Calculation and Basic Topology

To design the loop filter in Fig. 4, we must design a digital filter that operates at rate f_u and for which the closed-loop second-order PLL response has natural frequency ω_n and its damping factor ζ. This issue was discussed in depth in [8–10], where it was shown that the loop filter *B(z)* may be deduced from *F(s)* by using the *bilinear transformation method* [13 Section 7.1]. We now proceed to find *B(z)*.

$F(s) = K_a \frac{1+s\tau_2}{1+s\tau_1}$ has a pole at $s = -1/\tau_1$ and a zero at $s = -1/\tau_2$. To use the bilinear method, we pre-warp [13 Section 7.1]

$$\frac{1}{\tilde{\tau}_1} = \frac{1}{\pi T_u} \tan\left(\frac{\pi T_u}{\tau_1}\right) \text{and} \frac{1}{\tilde{\tau}_2} = \frac{1}{\pi T_u} \tan\left(\frac{\pi T_u}{\tau_2}\right) \tag{12}$$

and then construct the pre-warped transfer function of the analog filter $D(s) = K_a \frac{1+s\tilde{\tau}_2}{1+s\tilde{\tau}_1}$. Now we can employ the bilinear transformation [13 Section 7.1] in order to determine $B(z)$ as follows:

$$B(z) = D(s)\big|_{s=\frac{2}{T_u}\left(\frac{1-z^{-1}}{1+z^{-1}}\right)} = K_a \frac{1 + \frac{2}{T_u}\left(\frac{1-z^{-1}}{1+z^{-1}}\right)\tilde{\tau}_2}{1 + \frac{2}{T_u}\left(\frac{1-z^{-1}}{1+z^{-1}}\right)\tilde{\tau}_1}$$

$$= K_a \frac{\left(1+\frac{2\tilde{\tau}_2}{T_u}\right)}{\left(1+\frac{2\tilde{\tau}_1}{T_u}\right)} \cdot \left(\frac{1+\left[\left(1-\frac{2\tilde{\tau}_2}{T_u}\right)\Big/\left(1+\frac{2\tilde{\tau}_2}{T_u}\right)\right]z^{-1}}{1+\left[\left(1-\frac{2\tilde{\tau}_1}{T_u}\right)\Big/\left(1+\frac{2\tilde{\tau}_1}{T_u}\right)\right]z^{-1}}\right)$$

Now let us define $\beta_1 = (1-2\tilde{\tau}_2/T_u)\big/(1+2\tilde{\tau}_2/T_u)$, $\alpha_1 = (1-2\tilde{\tau}_1/T_u)\big/(1+2\tilde{\tau}_1/T_u)$, and also $\gamma = K_a(1+2\tilde{\tau}_2/T_u)\big/(1+2\tilde{\tau}_1/T_u)$, then we have

$$B(z) = \gamma \cdot \left(\frac{1+\beta_1 z^{-1}}{1+\alpha_1 z^{-1}}\right) \tag{14}$$

where

$$-1 < \beta_1 < 0, -1 < \alpha_1 < 0, \ \gamma > 0 \tag{15}$$

Moreover, it was shown that there exists a Direct-Form II implementation [13 Chapter 6] of $B(z)$, as shown in Fig. 6. Other filter topologies, such as the Direct-Form I topology [13 Chapter 6] are also possible, but it is easily seen that they are less efficient since they require more registers.

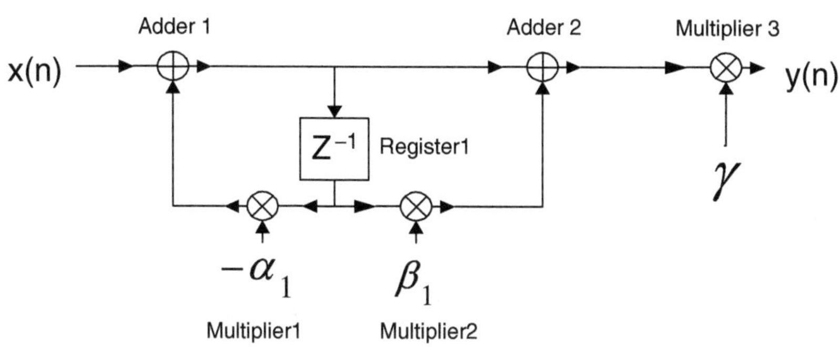

Fig. 6 Direct-Form II topology for $B(z)$

6 Improvement of Topology Through Pipelining

When implementing any structure inside FPGAs, an omnipresent desire is to design the structure with the shortest and simplest critical path as possible. Our loop filter needs to operate at rate f_u, and it is easily seen in Fig. 6 that critical path for the chosen topology is from the output of Register1, through Multiplier1, Adder1, Adder2, and Multiplier3.

The major problem with this critical path is the fact that it contains two multipliers. An improvement is therefore possible by adding a pipelining register, named Register2, between Adder2 and Multiplier3. This is shown in Fig. 7. As seen there, the critical path is now from the output of Register1, through Multiplier1, Adder1, Adder2, to the input of Register2. This critical path now contains only a single multiplier.

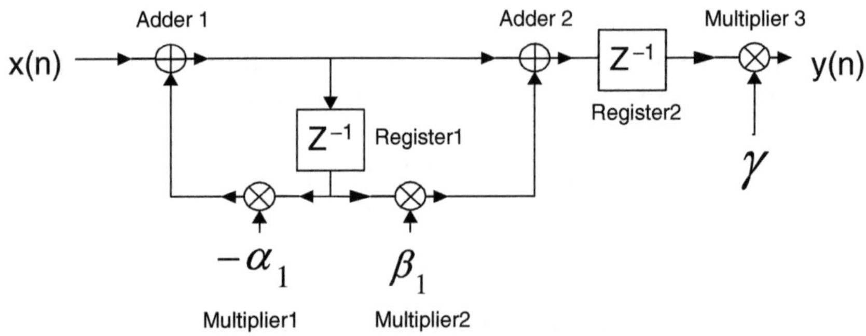

Fig. 7 Pipelined Direct-Form II topology for $B(z)$

Multipliers in current FPGAs, while much slower than adders, can generally operate quite fast and it might seem that reducing the critical path to contain only one multiplier has little practical advantage. However, later (in Sections 7–9) we shall see that this does indeed have great practical significance when we implement the multipliers as state machines.

One detrimental but generally minor side effect of the addition of the pipeline register in Fig. 7 is that it adds a delay element to the PLL whose delay is $T_u = 1/f_u$ (see [8 Fig. 22]). Such delays must be taken into account in PLL design insofar as they affect the PLL's phase margin (see [8 Section 10]).

7 Exact Implementational Parameters – A Case Study

In much of the literature digital filters are treated as mathematically abstract topologies where quantization and other implementation details are alluded to but rarely presented. Here we shall attempt to avoid this omission by discussing a case study of a specific loop filter structure that was implemented and tested by the author in the implementation of a 90 Mbps BPSK (binary phase shift keying) receiver. In that receiver the digital part was implemented in an FPGA using fixed-point arithmetic. Though a case study, the parameters and design choices discussed here may very well be applicable to many other systems with little modifications, due primarily to the fact that the overwhelming majority of PLLs are 2nd-order.

7.1 Binary Format

The binary representation is chosen as fixed-point signed two's complement format. Regarding the represented quantities, the presence or absence of a binary point is an arbitrary decision and does not affect the analysis (so long as such assumptions are made consistently). The chosen representations are (a) the filter coefficients β_1, α_1 are fractional (i.e., they have one sign bit and the rest of the bits represent a fraction); (b) the input and output of the filter, $x(n)$ and $y(n)$, respectively, are whole numbers; and (c) the coefficient γ has both whole and fractional parts (see Section 1.7.5).

7.2 Coefficient Quantization

$B(z)$ is an IIR (infinite impulse response) filter, and exact quantization analysis of such filters is in general complicated [13 Section 6.7.2]. However, because this filter has only one pole and one zero, it can be thought of as an extremely simple 1-stage cascade filter [13 Fig. 6.14]. Then, the data in [13 Section 6.8, Fig. 6.47] suggest that quantization of the coefficients to 16 bits is sufficient. Hence, this is the chosen quantization.

7.3 Input Quantization

The input quantization is chosen as 8 bits (i.e., 256 levels from − 128 to +127). This is justified as follows. The input quantization to the loop filter is equivalent to the output quantization of the phase detector. In the studied case (BPSK receiver) the phase detector is simply the decision-directed detector [1 Chapters 5, 6] $Q(n) \bullet sign(I(n))$ (where the sampling rate is 1 sample/symbol). Now, the precision of this phase detector is obviously that of the $Q(n)$ value, as sampled by the samplers. The number of bits of the sampler is thus a reasonable choice for the phase detector's output quantization. In the example system considered here, the samplers are 8-bit samplers (a common design choice), and hence the choice of 8-bit input quantization for the loop filter.

7.4 Overflow Considerations

The most problematic node in terms of overflow analysis is Adder1. To see this, it is advantageous to use the filter model shown in Fig. 8. This filter model is easily seen to be equivalent to Fig. 7. In Fig. 8, we see that $B(z)$ can be analyzed as a simple 1-stage IIR filter followed by an FIR (finite impulse response) filter. Now, if bus and register widths are properly chosen, then FIR filters will never overflow [13 Chapter 7]. On the other hand, the accumulator register (Register1) in the IIR filter will contain values that theoretically depend on a weighted sum of all of the previous

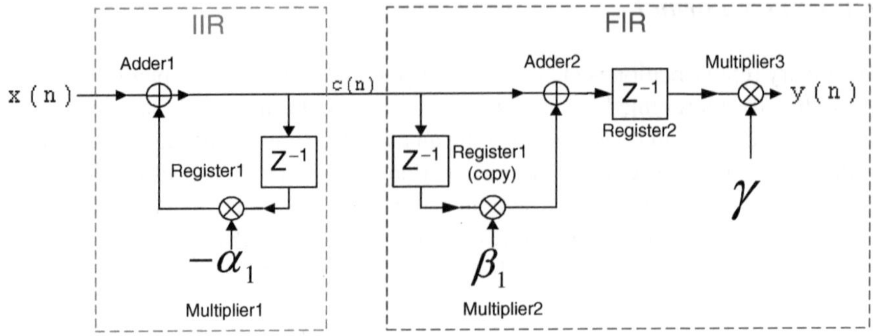

Fig. 8 Equivalent filter model suitable for overflow analysis

values of $x(n)$. Indeed, it is easily seen that for a PLL's loop filter we will have that $0 < -\alpha_1 < 1$ and, in fact, we will have that $-\alpha_1$ will be very close to 1, i.e., the IIR filter in Fig. 8 will be very close to an ideal integrator. Thus, an overflow may theoretically occur at the output of Adder1 if we are not careful (in the limiting case if $x(n)$ is constant any true [i.e. with $-\alpha_1 = 1$] integrator will, over time, overflow).

Fortunately, since $0 < -\alpha_1 < 1$, it is easy to avoid overflow by the following method. In Fig. 8 we have that $c(n) = x(n) - \alpha_1 c(n-1)$. Assume that we can design the filter so that the maximal value of $c(k)$ (for any k) is $c_{\max}$. It follows that

$$c(n) = x(n) - \alpha_1 c(n-1) \leq \max(x(n)) + \max(-\alpha_1) \cdot c_{\max} \tag{16}$$

Now, assume a worst case scenario where $-\alpha_1 = \max(-\alpha_1)$, $x(n) = \max(x(n))$, and $c(n-1) = c_{\max}$. In that case (16) will be an equality, i.e.

$$c(n) = \max(x(n)) + \max(-\alpha_1) \cdot c_{\max} \tag{17}$$

Equation (17) is the worst case scenario, so for $c_{\max}$ to exist we must have that $c(n) = c_{\max}$, and from (17)

$$c_{\max} = \frac{\max(x(n))}{1 - \max(-\alpha_1)} \tag{18}$$

Similarly,

$$c_{\min} = \min(x(n))/(1 - \max(-\alpha_1)) \tag{19}$$

If we design Register1 so that it contains enough bits to represent both $c_{\max}$ and $c_{\min}$, then we shall be assured that overflow never occurs. This will happen if

$$q = \lceil \log_2(\max(|c_{\min}|, |c_{\max}|)) \rceil + 1 \tag{20}$$

where q is the number of bits in Register1, $\lceil \bullet \rceil$ is "round up to the nearest integer," and the addition of 1 is due to the necessity for a sign bit.

The maximal value of $-\alpha_1$ in the signed two's complement 16-bit coefficient quantization will be $\max(-\alpha_1) = (2^{15} - 1)/2^{15} = 32767/32768 = 0.99997$. The input $x(n)$ is whole and quantized to 8 bits, so $\max(x(n)) = 127$ and $\min(x(n)) = -128$. Then, from (18), (19), and (20) we find that we need at least $q = 23$ bits (including sign bit) that represent a whole number in Register1.

To minimize effects due to quantization (and because it does not cost us much) we over-engineer Register1 to be a 32-bit register that represents a two's complement binary number composed of 1 sign bit, 23 whole bits, and 8 fractional bits. The extra bit added to the whole bit representation assures us that there is no overflow at the output of Adder2, since we have $|\beta_1| < 1$ so the absolute value of the output of Adder2 is at most $2(\max(|c_{\min}|, |c_{\max}|))$, and so adding another bit to the representation increases the dynamic range by a factor of 2 and assures that there is no overflow there.

7.5 Detailed Implementational Diagram

A diagram of the filter implementation that shows the bus widths is shown in Fig. 9. In that figure, we adopt the following notations: s = sign bit, w = bits that are part of the representation of the whole part of the number, f = bits that are part of the representation of the fractional part of the number, and e = sign extension bits (i.e., bits that mathematically will always be equal to the sign bit).

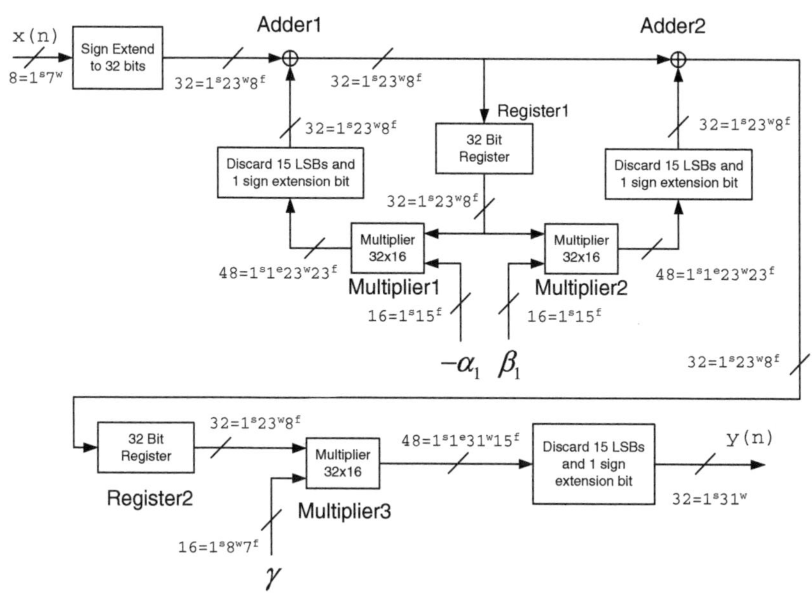

Fig. 9 Detailed filter implementation showing bus widths

8 Improvement of Logic Resource Utilization via Innovative Multiplier Implementation

In Fig. 9 the multipliers are by far the costliest elements in terms of logic resource requirements. In this section we show that it is possible to achieve extraordinary savings in logic resource requirements through an innovative implementation of these multipliers.

8.1 The Basic Idea

To initiate this discussion, we first discuss how multipliers are implemented. Consider the multiplication of two numbers, say 83 and 57. Multiplication is done[2] by shifting and adding the partial products, as shown in Fig. 10.

The conventional multiplier implementation is oriented toward achieving minimal latency and as such computes all the partial products in parallel. However, in the case discussed in this chapter we can use (3) to make the following observation. If the loop filter clock rate is made slow enough, then we can use a *state machine* to compute the partial products in *sequence* rather than in parallel. By iteratively shifting and adding these partial products we will thus arrive at the desired result. From an efficiency standpoint, it is advantageous to start with the partial product of the MSB (most significant bit) and then consecutively shift left by 1 bit and add the partial products of each lower bit until the LSB (least significant bit). Now, each partial product in binary is basically[2] multiplication either by 1 (simply the other number) or by 0 (which is 0). Therefore, the state machine itself will not need any multipliers. Hence, there is potential here for great savings in logic resources.

```
                                            1010011
                                          × 111001
                     83                     1010011
  In decimal:       ×57    or in binary:   0000000
                    581                   0000000
                   415                   1010011
                   4731                 1010011
                                       1010011
                                       1001001111011
```

Fig. 10 Multiplication of the numbers 83 and 57

[2] Here for simplicity we are multiplying two positive numbers. When one or more of the numbers is negative then tricky sign and sign-extension issues are present. These issues are quite easy and straightforward to resolve, and this subject is treated in Section 8.2.1.

8.2 State Machine Algorithm and Implementation

To implement a state machine that multiplies a 32-bit number by a 16-bit number (as is needed in Fig. 9) then to reduce the number of state machine clock cycles needed to compute the multiplication it is advantageous to implement a machine that sums fifteen 31-bit partial products rather than one which sums thirty-one 15-bit partial products (the sign bits are excluded from partial product computations)[2].

8.2.1 State Machine Algorithm for the Multiplier

A simplified flowchart of the state machine's algorithm is shown in Fig. 11 (note that this is a conceptual flowchart and the boxes do not necessarily each correspond to a state). In Fig. 11 we assumed that the first multiplicand `A` is a 16-bit number and the second multiplicand `B` is a 32-bit number. The result is given in the variable `Result`

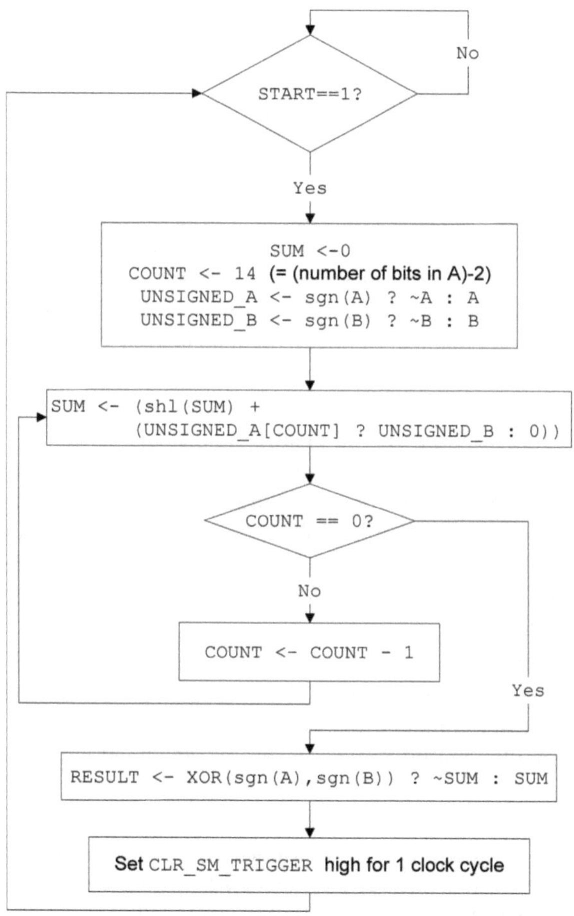

Fig. 11 Simplified flowchart of multiplier state machine algorithm. The multiplier multiplies `A[15:0]` by `B[31:0]` and outputs `Result[47:0]`. All quantities are in two's complement notation. Some notations used are "~" is bit-wise negation; "`sgn(x)`" means the sign of x, i.e., the MSB (most significant bit) of x; "`shl`" means shift left by 1 bit. The syntax "`y<-x ? a : b`" is shorthand for "`if(x==1)then y<-a else y<-b`"

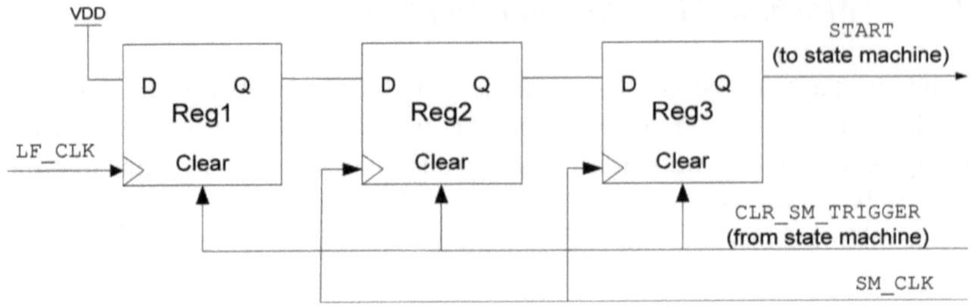

Fig. 12 Generation of start strobe to the multiplier state machines. `Lf_Clk` is the loop filter clock. `Sm_Clk` is the state machine clock. `Vdd` is the logical "1" voltage.

which is a 48-bit number. For the meaning of `Start` and `Clr_Sm_Trigger`, see Section 8.4 and Fig. 12.

As can be seen in Fig. 11, we solve the issue posed by signed data by first multiplying the unsigned data and then adjusting the result according to the correct sign. This is best explained in its decimal analogy. To multiply 83 by (–57), we can multiply 83 by 57 (achieving 4731) and then negate the result (thus arriving at the correct result of –4731).

However, the sharp-eyed reader will have noticed that the multiplication algorithm shown in Fig. 11 has some mathematical flaws. The problem in Fig. 11 is that in two′s complement arithmetic, bitwise negation does not correspond to the negative of the number. Rather, in two′s complement arithmetic the negative of a number is achieved by bitwise negation followed by addition of 1. Therefore whenever in Fig. 11 it is written "`~x` " it should be written "`(~x)+1`."

There are two questions that deserve an answer: (a) Why is the state machine implemented as in Fig. 11, and why is this implementation actually preferable to the mathematically correct implementation? and (b) Why are such mathematical inaccuracies permissible in our system?

The answer to question (a) lies in a quirk of two′s complement arithmetic, which is that its negative and positive ranges are unequal. For example, in 16-bit two′s arithmetic we can represent numbers from $-32768 (=8000_{16})$ to $32767\ (=7FFF_{16})$. Therefore, if we were to accurately negate -32768, we would have to represent 32768, which is impossible to do in a 16-bit two′s complement number. Thus, by eliminating the "+1" stage of the negation, we eliminate the very serious problem of potential overflow. Obviously, we also achieve a reduction in logic resources by not implementing the addition.

The answer to question (b) is slightly more thought-provoking. In the previous paragraph we have shown that our non-standard method of negating numbers results in an inaccuracy in the order of magnitude of the LSB of the result of the negation. On their own, these errors are small and negligible[3]. But can these errors not

[3] The fact that we have over-engineered the filter by adding eight more LSB bits to the datapath to represent the fractional part of the results of operations (see Fig. 9, Section 7.4) also helps since this reduces the error magnitude of an LSB error by 256, i.e., more than 2 orders of magnitude.

accumulate? The answer is no. This is because we must remember that the multipliers operate within a PLL, i.e., a closed-loop *feedback* system. The small mathematical errors in the loop-filter's output will thus be corrected by the PLL's normal feedback operation. See also [14 Section IV-A] for discussion of a similar situation.

Thus, the small sacrifice in mathematical correctness is irrelevant for the current application, but the chosen imprecise implementation affords logic savings and the inherent avoidance of overflow problems. The reader is advised, however, to apply extreme caution when thinking of using the algorithm Fig. 11 in other settings, especially in an open-loop system.

8.2.2 Calculation of the Required State Machine Clock Rate

Assume (as was indeed the case in the example receiver under discussion) that we have designed the multiplier state machine so that it takes 2 clock cycles per partial product, as follows:

- (1st clock cycle): Shift the accumulated sum of previous partial products by 1 bit to the left;
- (2nd clock cycle): Sum to this accumulated sum the partial product corresponding to the current bit;

then it will take $15 \times 2 = 30$ clock cycles to shift and sum 15 partial products (each 31 bits long [note, again, that sign bits are excluded from this process since we operate on the unsigned data, see Section 8.2]). If we allow for an additional 5 clock cycles for the state machine to start and finish and other overhead, we arrive at 35 clock cycles. Obviously, the engineer must also allow time for the adders in the loop filter to process the results of the multiplier before the next f_u clock edge (the critical path being from the output of Multiplier1 through Adder1 and Adder2), as well as allow time for the setup times of the registers to be complied with. However, those latencies are usually small and are easily modeled by FPGA design software, so their inclusion in the calculations is easy. Another source of latency is caused by the necessity to synchronize the start strobe of the state machine (see Section 8.4). For the purposes of the example in this chapter, we round the figure 35 clock cycles to 40 clock cycle for good measure (in order to achieve extra "engineering robustness" and to take into account the aforementioned additional latencies). This means that if the rate f_u is 40 times slower than the state machine clock, then we can compute the multiplications using the state machine and the results will propagate through the loop filter's combinational logic paths before the next loop filter clock edge arrives.

Determination of f_u is a subject that is studied in depth in [8–10]. To give an example, if we want to design a PLL with $f_n = 2000$ Hz, then good results can be obtained if $f_u = 700{,}000$ Hz. Thus, to implement the multipliers as state machines, for this case we will need a state machine clock of at least $40 f_u = 28$ MHz, which is quite a reasonable state machine clock.

8.3 The Importance of the Pipeline Register

Now is a good time to make a note of the importance of the pipeline register Register2. When the multipliers were implemented as fast modules where the partial products were computed and summed in parallel, then Register2 afforded little to no advantage. However, now that we are using a state machine, the fact that the critical path contains only one multiplier (instead of two) allows us to use a slow state-machine clock. For example, without Register2 the critical path in the example of the previous subsection would contain Multiplier1 and Multiplier3 and both would be required to finish their computation – in series – within T_u seconds (and also allow time for other latencies as mentioned in Section 8.2). This would result in a required state machine clock rate that is about twice as fast, i.e., about 80 $f_u = 56$ MHz.

8.4 Triggering of the State Machine

Since the loop filter clock with rate f_u is in general not synchronized to the state machine clock, it is necessary to find a way in which to trigger the state machine's operation. This is done using the structure shown in Fig. 12. As seen there, the rising edge of the loop filter clock will cause a "1" to propagate through two registers which are clocked by the state machine clock. The resulting signal (denoted as `Start`) can serve as the start input to the state machine and is synchronized to the state machine clock. The two registers are necessary in order to avoid metastable effects [15 Section 10.3.3] during synchronization of the start strobe, and the delay incurred as a result (worst case: 3 `Sm_Clk` clock cycles) must be taken into account when computing the required state machine clock (see Section 8.2). After the multiplication is completed, the state machine sends a clear signal (denoted as `Clr_Sm_Trigger` in Fig. 12) to the registers which readies them for the triggering of the next multiplication.

9 Quantitative Logic Resource Savings Results

In order to quantitatively evaluate the benefits of the proposed implementation, hardware mapping of two loop filter implementations was done on a Xilinx Virtex XCV600-4HQ240 chip [16]. The design software used was Xilinx ISE 8.2.03i.

The first loop filter implementation contains "conventional" multipliers implemented using the Xilinx Core Generator, which results in extremely logic-efficient implementation. These multipliers used the Xilinx's multiplier version 8.0 core and used the most resource-efficient implementation, that is, a non-pipelined implementation (i.e., combinatorial logic only).

The second loop filter implementation contains multipliers implemented as state machines, as outlined in Section 8.2. The results of the comparison are shown in

Table 1. There are various ways to measure resource utilization in FPGAs. In Table 1 we present two metrics: the total equivalent gate count and the number of occupied FPGA slices. The results show that the proposed implementation method results in a logic resource savings of between 71 and 76%.

Table 1 Hardware mapping comparison using the Xilinx Virtex XCV600-4HQ240 chip. The results are for the *entire* loop filter (not just the multipliers)

Multiplier implementation	Total equivalent gate count	No. of occupied slices
"Conventional"	22,085	855
state machine	6,408	206
Resource savings	71%	76%

10 Conclusions

In this chapter we discussed the design of digital loop filters for phase lock loops in high-speed wireless receivers. It was found that if certain conditions regarding the phase detector sample rate and the PLL′s natural frequency are fulfilled, then significant savings in resource utilization (between 71 and 76% in the example presented) can be achieved. The reduction in resource usage was accomplished by implementing the multipliers within the loop filter as state machines which compute and sum the partial products iteratively, rather than via a conventional multiplier that computes and sums the partial products in parallel. It was further found that a modified Direct-Form II structure in which a strategically placed pipeline register is inserted is a suitable filter structure for this type of multiplier implementation. The method proposed in this chapter has been used by the author in the implementation of a 90 Mbps BPSK receiver where the digital portion of the receiver was implemented in a Xilinx Virtex XCV1000-6BG560C chip, and the parameters of the carrier synchronization PLL of that system were investigated as a case study in this chapter. Moreover, it shall be commented that in that receiver, loop filters for various PLLs and control loops were implemented using the proposed technique, including loop filters for the carrier PLL, the symbol timing synchronization PLL, and two AGC (automatic gain control) loops. Indeed, in the aforementioned system, the implementation of the loop filters using the efficient method presented here was crucial in order to allow the entire receiver design to fit in one single FPGA. Thus, the proposed design method has been proven in practice and can be a valuable tool for the implementation of contemporary receivers.

Acknowledgment The author gratefully acknowledges the financial support provided by NSERC (National Sciences and Engineering Research Council of Canada) through its Canadian Graduate Scholarship.

References

1. H. Meyr, M. Moeneclaey, and S. Fechtel, *Digital communication receivers: synchronization, channel estimation, and signal processing*. NY: Wiley, 1998.
2. U. Mengali and A. N. D'Andrea, *Synchronization techniques for digital receivers*. NY: Plenum Press, 1997.
3. H. Meyr and G. Ascheid, *Synchronization in digital communications*. NY: Wiley, 1990.
4. F. M. Gardner, *Phaselock techniques*, 2nd ed. NY: Wiley, 1979.
5. R. E. Best, *Phase-locked loops: theory, design, and applications*, 2nd ed. NY: McGraw-Hill, 1993.
6. W. P. Robins, *Phase noise in signal sources. (Theory and applications)*. London: Peter Peregrinus, 1982.
7. ETSI (European Telecommunications Standards Institute), "DVB-S2 Technical Report ETSI TR 102 376 V1.1.1," 2005.
8. Y. Linn, "A Methodical Approach to Hybrid PLL Design for High-Speed Wireless Communications," in *Proc. 8th IEEE Wireless and Microwave Technology Conf. (WAMICON 2006)*, Clearwater, FL, Dec. 4–5, 2006.
9. Y. Linn, "A Tutorial on Hybrid PLL Design for Synchronization in Wireless Receivers," in *Proc. International Seminar: 15 Years of Electronic Engineering*, Universidad Pontificia Bolivariana, Bucaramanga, Colombia, Aug. 15–19, 2006 *(invited paper)*.
10. Y. Linn, "Synchronization and Receiver Structures in Digital Wireless Communications (workshop notes)," in *International Seminar: 15 Years of Electronic Engineering*. Universidad Pontificia Bolivariana, Bucaramanga, Colombia, Aug. 15–19, 2006.
11. A. Blanchard, *Phase-locked loops. Application to coherent receiver design*. NY: Wiley, 1976.
12. J. J. D′Azzo and C. H. Houpis, *Linear control system analysis and design: conventional and modern*, 3rd ed. NY: McGraw-Hill, 1988.
13. A. V. Oppenheim and R. W. Schafer, *Discrete-time signal processing*. NJ: Prentice Hall, 1989.
14. F. M. Gardner, "Interpolation in digital modems. I. Fundamentals," *IEEE Trans. Commun.*, 41, 3, 501–507, Mar. 1993.
15. S. Brown and Z. Vranesic, *Fundamentals of Digital Logic with VHDL Design*, 2nd ed. NY: McGraw-Hill, 2005.
16. Xilinx Inc., "Virtex Series FPGAs," at http://www.xilinx.com/products/silicon_solutions/fpgas/virtex/index.htm, accessed Nov. 2006

Finite Automata for Evaluating Testbed Resource Contention

Lei Liu

Abstract Next-generation Internet and wireless telecommunication networks compose a pervasive computing structure. Real large-scale tests, simulation and emulation are the major strategies to construct next generation testbed. Synchronization primitives are critical to ensure finite resource assesses. For performance load and utilization prediction, this research proposes a formal testbed abstraction and contention model grounded from automata theory. To conduct a configuration analysis, algorithm operations are exploited. In addition, an empirical testbed with self-organization, dynamic resource allocation, partition, virtualization, and scheduling is deliberated.

1 Introduction

Next-generation Internet and wireless telecommunication networks suggest the setting of a pervasive computing graph abstracted with a set of large-scale distributed systems, small and resource constrained devices, and communication edges or links. Nodes share the same physical media. The data-link layer manages link resources and coordinates medium access among peers. The network layer maintains communication paths. Mobility and the degree of adaptation may vary from application awareness to agnostic.

The properties of the graph hold a partially observable and stochastic task environment due to incomplete environmental data collection, input, and uncertainty of future states. In addition, it is a dynamic and sequential task, since a current load can influence test performance and consequently have impact on future testbed sizing and capacity planning. Moreover, divergent vendor-specific architectures and implementations pose a challenge to a finite experimental environment in

L. Liu (✉)
3738 Evangelho Cir., Sane Jose, CA 85149, USA
e-mail: lei.liu@sun.com

S. Powell, J.P. Shim (eds.), *Wireless Technology*, Lecture Notes in Electrical Engineering 44, DOI 10.1007/978-0-387-71787-6_9,

performance instrumentation. An infrastructure for the above fixture is referred to as testbed.

Testbed strategy could be a combination of full-scale tests, simulation, and emulation. Exploratory actions with real test environment such as production probing and tracing are limited by the period of executions and overheads of management traffics. In addition, implementation may involve system level programming and deployment of large-scale testbed. However, scientific methods such as simulation models [4] require low-level and development of programs and validation against a probabilistic distribution of job or task workloads. Emulation [5] provides a middle ground between the above simulation and full-scale experimental capabilities with reproducible results.

The problem of architectural issues of developing a scalable Internet and wireless testbed is to synchronize access to shared computational resources such as vertices, edges, and links using locking primitives and techniques. A lock is a single byte location in RAM. It has two mutual exclusive states as free (**0x00**) or acquired (**0xFF**). Implementing a locking scheme that only does one or the other can severely affect scalability and performance. The number and type of locks need to be deterministic and comply with the lock hierarchy and rules of acquiring locks toward resource usage and data manipulation. Hence, contention, the time spent to access resources is important to estimate total response time.

2 Main Thrust of the Chapter

Next generation testbed is different from conventional classes of concurrent environmental model in terms of uncertainty and independence with dynamics of processors, topologies, input, and locality. Within the proceeding environmental setting, a set of computing resources was located at nodes within a network graph:

$$\mathbf{G} = (\mathbf{V},\mathbf{E}) \tag{1}$$

A vertex is a virtual environment identified within the directed graph, which is a bookkeeping data structure. An edge is an attack path or link to another vertex. The edge between a pair of vertices **v** and **u** will be (**u**, **v**), where **u** and **v** belong to **V**. A testbed is denoted to represent a finite set of processes for targeted resources.

Finite automaton [7] is a useful model for many important kinds of hardware and software to verify systems that have a finite number of distinct states. Next generation testbed is composed of many systems or components and viewed as being at all times in one of finite number of states. It is further formulated as a finite automaton with a set of execution states of test plans with state transactions in response to input $\boldsymbol{\Sigma}$ originated from external actions such as load generation and administrative commands. Self-organization and resource allocation are considered as internal actions, which may affect state changes. The final states of testbed automaton are denoted as a set **F**.

2.1 Problem Formulation

A testbed automata (shown in Fig. 1) is defined as

$$\mathbf{A_{testbed}} = (\mathbf{Q}, \mathbf{\Sigma}, \boldsymbol{\delta}, \mathbf{q}_0, \mathbf{F}) \tag{2}$$

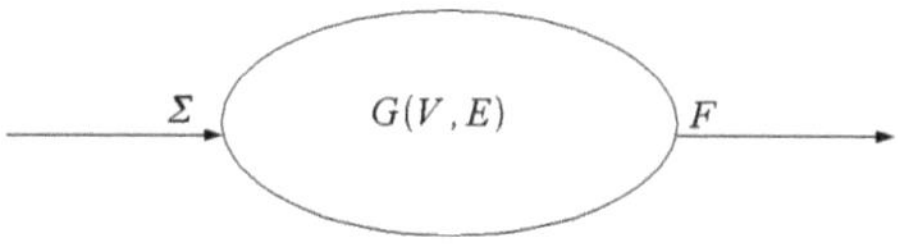

Fig. 1 Testbed automaton

States: Each state represents a situation that testbed could be in. The state is to abstract the specific important events, which have taken place as a sequence of actions, whereas others not yet happened. In addition, the state is also to record the relevant portion of testbed history. Since testbed automata only have a finite number of states and the pervasive state history are generally cannot be stored, so next generation testbed must be designed carefully to remember what is important and neglect the unimportant. Moreover, the major advantage of adopting a finite number of states is to implement testbed with a finite set of automata elements and computational resources. The state container is specified as a structure with the following properties:

- The set of vertices existing in $\mathbf{A}_{\mathbf{testbed}}$:
- The set of channels (edges and links) existing in $\mathbf{A}_{\mathbf{testbed}}$.
- The set of lock automata existing in $\mathbf{A}_{\mathbf{testbed}}$.

Initial State: For simplicity, testbed is constructed with physical vertices, edges, and links but without initial contention property. In each state management of testbed automata, both external state and internal controls organize the computational resources before the execution of next automata input:

$$\mathbf{vertices}[\mathbf{A_{testbed}}] \leftarrow \mathrm{N} \tag{3}$$

$$\mathbf{channels}[\mathbf{A_{testbed}}] \leftarrow \mathrm{N} \tag{4}$$

Transition Function $\boldsymbol{\delta}(\mathbf{q}, \mathbf{a})$: As state transition illustrated in Fig. 2, transaction functions are internal controls to allocate testbed computation resources. However, even for self-organized testbed, external interactive administration controls could be automata input to justify the state manipulation.

Input $\mathbf{\Sigma}$: Testbed automata input could be quantified as deployment of test matrix, simulation packages, and emulation software artifacts.

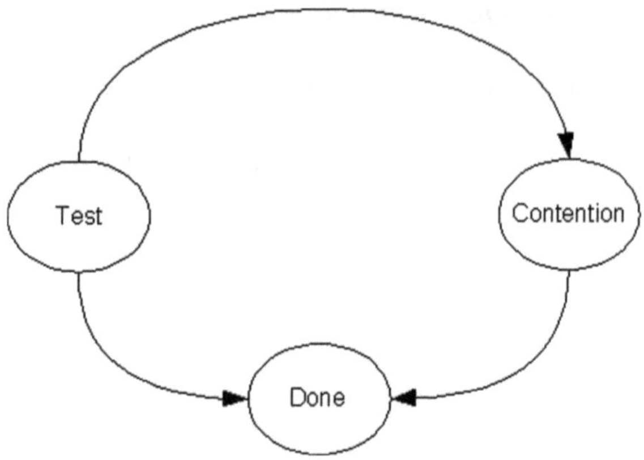

Fig. 2 Testbed automaton state transition

Output **F** *and Goal Test*: The success of state transitions is measured upon the fairness of execution of test plans, contention reduction, and responsiveness of test results.

Cost: If the bound of test duration is the step time $\mathbf{O(t_i)}$ then the time measurement of testbed automata complexity is the discrete summation of $\mathbf{a_k}$as the time assigned to a specific input $\mathbf{\Sigma}$.

This process is one of the most fundamental abstractions provided by testbed automaton. It is an instance of program running on each vertex. Since each process is an executable form of file associated with exclusive resource utilization within testbed, resources required by processes are abstracted for execution time on processors and allocation of physical memory, as well as to perform authorized services, such as network and disk I/O. Specifically, within the same instance of operating system, a process state can be maintained at kernel address space for the occupied pages of physical memory, which hold specific memory segments with computational components such as instructions, stack space, data space, and other necessities for execution. Therefore, a concrete formulation of testbed automaton is a summation of finite process automaton. Hence, a contention automata is defined as

$$\mathbf{A} = (\mathbf{Q},\mathbf{\Sigma},\boldsymbol{\delta},\mathbf{q}_0,\mathbf{F}) \tag{5}$$

States **Q**: Each contention state represents a situation that locks could exist. Similarly, the state indicates that specific contention events have taken place. The contention state is designated to record the every critical portion of contention history. Since there are only a finite number of contention states, the entire contention histogram is generally not necessary, so the contention automata must store what is important and neglect what is not. The major advantages of adopting a finite number of states are that we can implement the system with a finite set of resources. The properties of a contention state are defined as

- the number of lock contentions that exist in;
- the duration of sleep time for blocking contentions; and
- the number of spin counts for non-blocking contentions.

Initial State $\mathbf{q_0}$: At any given time the number of locks is dynamic. Locks can be created dynamically during normal operations. For instance, kernel threads and processes are created, file systems are mounted, files are created and opened, and network connections are made. The starting state is init state. To support concurrent test plans, multiple initial states could exist in automaton A.

Transition Function $\boldsymbol{\delta}(\mathbf{q}, \mathbf{a})$: As state transition illustrated in Fig. 3, transaction functions are internal controls to allocate testbed computation resources.

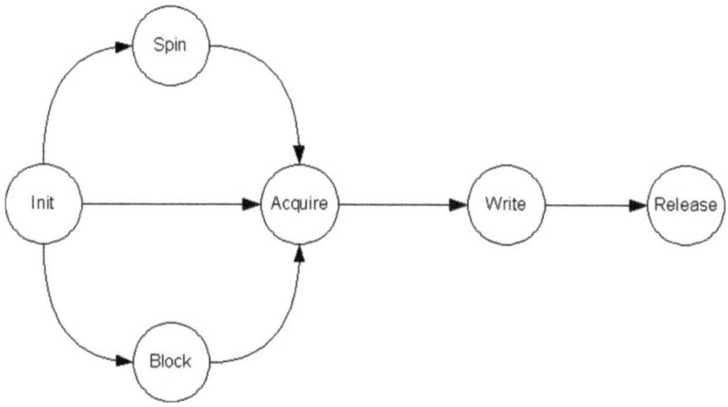

Fig. 3 Contention automaton state transition

The internal controls are managed by kernel scheduler or generic resource management daemon processes as automata input to manipulate contention state transitions.

Input $\boldsymbol{\Sigma}$: Contention automata input is the source of data information resulting in resource contentions.

Output **F** *and Goal Test*: The success of state transitions is measured upon the fairness of execution of contented processes or kernel execution threads. The releases of lock events are the exiting stages of lock contentions.

Cost: If the bound of a contention resolution is the step time $\mathbf{O(t_i)}$ then the time measurement of contention automata complexity is the discrete summation of $\mathbf{a_k}$ as the time assigned to a specific lock hold event from contention occurrence to the releases of locks.

2.2 Contribution

The purpose of this research is to provide an apparatus and method for a resource organization within next generation Internet or wireless network infrastructure and associative task environment. To construct real large-scale tests, simulation, and emulation tests within next generation test environment, one needs to depict and reason about capacity and configuration of vertices, edges, and links within testbed computational graph. Deterministic finite automata (FDA) theory [7] is applied. Specifically, testbed and contention automata are suggested. The automata models are proposed to simplify the resource organization and prediction. To explore the interactions of automata, a product automaton is formed to represent contention within concurrent test environment. In addition, a set of testbed algorithmic operations is proposed to resolve lock contentions and self-organization of resource allocation. Furthermore, an empirical testbed is advocated with self-organization in terms of resource partition, allocation, constraints, and virtualization for public access.

2.2.1 Review of Literature

The main threads of conventional testbed researches fell into three categories: real tests, simulation, and emulation. The most influential testbed strategy was simulation-focused study [4]. Simulation requires less physical resources involved in the small-scale test. However, experiment has dependency with simulation test input distribution. This distribution is normally built upon design and test hypothetical assumptions about traffic, failure patterns, and topology.

Another thread of testbed research is to expose experiments to real task environment. Real-world test environment ideally was proposed to construct protocol evaluation [6] for ad hoc networks. The techniques are designated to achieve scalable and reproducible experiments over large-scale networks. However, since real large-scale tests require distributed design and implementation other than massive deployment and configuration with the complexity of nodes, transport protocols, middleware, and applications. Moreover, real experiments with distributed sensor networks could become troublesome if the nodes grow and exceed a dozens. Self-organization of testbed remained as an open research issue. Hence, it was limited to large carrier practices. Among the above two approaches, emulation [5] was in the middle. Emulation increases the experimental realism with acceptable reproducible results. Another thread of study [3] on testbed was initially discussed on the combination of the above three test strategies within distribution task environment. The study suggested adaptation to wireless networks from layer 2 to layer 7 to provide execution framework for both remote and local end users. This research augments the hybrid test strategies to propose self-organized testbed with adaptation, virtualization, partitioning, and dynamic resource allocation.

From automata literature, an extension of the probabilistic I/O automata framework [1] has been studied to support modeling and verification of security protocols. However, there is no existing discussion on automata and related resource contention within testbed state space. This research proposes testbed and contention automata for experiments within next generation Internet and wireless networks.

2.2.2 Methodology

As a generalized view of next generation testbed automaton, the detailed implementation and deployment aspects are abstracted. To reduce complexity, a general goal-based finite automaton model is developed in order to abstract out some complexities of platform and software implementation of real tests while retaining those essential to meet the service objectives.

To map user land application execution into kernel scheduling and execution unit with associated physical resource utilization, one can abstract the complexity of an appliance system as three interconnected blocking and spin locks of operating system kernel resources and underlining hardware processors. The automaton model can be applied to analyze a workload density or a set of discrete input events with a given transaction arrival P of service data or content traffics. The instance of automaton at kernel space can abstract out physics of contention elements such as disk, network, and memory contributions to the potential performance degradation of good-put within testbed environment.

In addition, resource partitioning and isolation of execution environments could share a single instance of hosting operating system and multiple instances of guest operating systems with associated kernel resources. Kernel tunable, kernel built-in resource management module, and networking OSI reference implementation provide congestion and administrative controls on the bounds of resource utilization. Hence, both user land processes and kernel threads are explicitly considered as automata elements. Pertaining to blocking primitives, fair scheduling, priority scheduling, FIFO scheduling, and lock implementation are generalized as a generic queue, which abstracts the detailed queue implementation such as the direction of the queue for inbound and outbound packets and number of queues, etc.

For mutual exclusion, lock primitives ensure concurrency as a thread attempts to acquire a mutex lock that is being owned by another thread. The calling thread can either spin or block to wait for available resources. A spinning thread enters a tight loop in order to acquire the lock in each pass of the algorithmic operation. On the other hand, a blocked thread is placed on a sleep queue as another thread holds the intended lock. The sleeping threads will wakeup when the lock is released. The blocking locks require context switching to get the waiting threads off the processors and a new runnable thread onto CPUs. In addition, there is also a little more lock acquisition latency.

There are spin locks and adaptive locks for most of operating system implementation. Adaptive locks are the most common type of lock used and are designed to dynamically either spin or block when a lock is being held, depending on the state of the holder. Selection of a locking scheme that only does one or the other can severely affect scalability and performance. Only spin locks can be used in high-level interrupt handlers. In addition, spin locks can raise the interrupt level of the processor when the lock is acquired.

Other than kernel space generalization, user land activities need to be abstracted. Specifically, tested application specific lightweight processes, software threading model, and scheduling implementation are included in the automaton qualification in order to simplify the abstraction. A process is stopped, either by a job control signal or because it is being traced. Internal actions are the self-organization of testbed automaton (see Fig. 3).

To observe the behavior of automaton state transitions, event driven probe algorithm described below is used to discover the asynchronous subsequences, consisting of all the actions.

```
PROBE-CONTENTION(event, no, time)                                    (6)
 if(type[event] == USER-LAND)
    case BLOCK
      spins←no;
      lock++;
    case SPIN
      spins←no;
      lock++;
    case READ-WRITE-BLOCK
      sleep←time;
      lock++;
 end if
 if(type[event] == KERNEL)
    case ADAPTIVE-SPIN
      spins←no;
      lock++;
    case ADAPTIVE-BLOCK
      spins←no;
      lock++;
    case SPIN-SPIN
      spins←no;
      lock++;
    case READ-WRITE-BLOCK
      sleep←time;
      lock++;
 end if
```

2.3 Experimental Results

The testbed architecture was composed of a VLAN configuration. A single point of dual interface is configured to act as an Internet gateway. Within this private network, Sun N1^{TM} Grid Engine provides a uniformed job submission and scheduling service interfaces for test plans execution within testbed. In addition, grid engine has public accessible web user interfaces for job submission via HTTP. Sun N1 Service Provisioning System evaluates the proposed performance model. There was a Gigabit switch providing point-to-point connection between the preventive appliances with a load generator. The software packages were deployed on four SunFire™ T2000 servers with 32 x 1 Ghz Ultra-SPARC T1™ processors, 8 GB RAM, and 2 x 68 GB disks, and 4 x Gigabit Ethernet ports.

The T2000 servers were optimized with CMT and Ultra-SPARC T1 technology, firmware, and hardware cache design. In contrast to traditional processor design, which focuses on single hardware thread execution, T1 processor provides instruction level parallelism instead of thread level parallelization. It has multiple physical instruction execution pipelines and several active thread contexts per pipeline. In addition, improved hardware design with masking memory access reduces memory latency for processor spending most of its time stalled and waiting for memory. There is a 12 way associative unified L2 on chip cache. Double Data Rate 2 memory reduces stall. Furthermore, each board has switch chips to connect to on-board components.

Even more, Solaris™ 10 operating system was configured with optimal tuning parameters from kernel core modules to device drivers. Three key parameters of IP module needed to be specified to dispatch the worker threads to different processors (ip soft rings cnt = 32, ip sequence bound = 0, ip sequence fanout = 1) to execute the interrupt handlers in order to increase processor utilization and throughput. The critical TCP module tunable (tcp conn req max q, tcp conn req max q0) and the backlog of the user land process were set to 8 K to reduce the error rate and improve throughput. At user land, both listening backlog and listener threads are configured as optimal values. Service provisioning plan for resource pool configuration is listed as a segment of script as shown below.

```
pool testbed resource pool
  boolean pool.default false                                    (7)
  boolean pool.active true
  int pool.importance 1
  string pool.comment
  string pool.scheduler FSS
pset batch
  pset testbed pset
  int pset.sys id 1
  string pset.units population
  boolean pset.default true
  uint pset.min 2
  uint pset.max 4
  string pset.comment
  boolean pset.escapable false
  uint pset.load 0
  uint pset.size 0
  string pset.poold.objectives  locality tight;utilization<40
cpu
  int cpu.sys id 5
```

```
  string cpu.comment
  string cpu.status on-line
cpu
  int cpu.sys id 7
  string cpu.comment
  string cpu.status on-line
```

2.4 Implementation

Zone configurations are defined to implement resource partitioning. Resource pools are specified for dynamic resource allocation with constraints. Hardware virtualization provides heterogeneous test environment with both SPARC and x86 platforms. A public grid scheduler is enabled from web interfaces. Dynamic tracing tasks are scripted with D Language [2] to probe contention events at runtime as shown below.

```
PROBE-USERLAND( )                                                    (8)
  plockstat:::mutex-block
{
    lock = arg0;
    @locks["mutex-block"] = count();
}
plockstat:::mutex-spin
{
    lock = arg0;
    @locks["mutex-spin"] = count();
}
plockstat:::rw-block
{
    lock = arg0;
    @locks["rw-block"] = count();
}
PROBE-KERNEL( )
lockstat:::adaptive-spin
{
    spins = arg1;
    @locks["adaptive-spin"] = count();
}
lockstat:::adaptive-block
{
    sleep = arg1;
    @locks["adaptive-block"] = count();
}
```

```
lockstat:::spin-spin
{
    spins = arg1;
    @locks["spin-spin"] = count();
}
lockstat:::rw-block
{
    sleep = arg1;
    @locks["r-w-block"] = count();
}
```

For adaptive-block event, probe fires after a thread, which has blocked on a held adaptive mutex lock, has reawakened and acquired the mutex lock. Either adaptive-block or adaptive-spin will fire for a lock acquisition. For adaptive-spin event, probe invoked has successfully acquired the mutex as a thread that has spun on a held adaptive mutex. Spin–spin contention-event probe is triggered to acquire the spin lock after a thread that has spun on a held spin lock. Read-write contention events fire as reawakened and has acquired the lock soon after a thread that has blocked on a held readers/writer lock.

A snapshot (see Table 1) of Java^{TM} Virtual Machine (JVM) Lock contention was resulted from a web service integration sever. It shows the top eight calls at user land, which led to CPU contention within testbed.

Table 1 CPU contention samples

Rank	Self	Accum	Count	Function
1	31.04%	31.04%	1,713	DataFactory.create
2	30.14%	61.18%	1,663	PlainSocketImpl.socketAccept
3	12.25%	73.43%	676	TestGetDocument.getDocument
4	6.80%	80.23%	375	createReader
5	3.01%	83.24%	166	Thread.setPriority0
6	2.90%	86.14%	160	RecordToDocumentService.recursive ToDocument
7	2.08%	88.22%	115	lookup
8	1.72%	89.94%	95	getCursor

2.5 *Future Trends*

Dynamic tracing, resource allocation, partitioning, virtualization, self-healing, and kernel resource constraints are critical characteristics to provide self-organization to next generation testbed. Simulation, emulation, and real large-scale test plan can be achieved with simplification. Future work is to conduct a further study on the semantics of testbed and contention automata for both cooperative and competitive tasks within large state spaces with high dimensionality.

2.6 Biography

Lei serves as a senior engineer for Sun Microsystems, Inc. He has about 10 years of experience on full life cycle R&D, product development, integration, testing, and partner engineering engagement. Lei has accomplished various tasks from latest hardware CMT platform, and Grid Rack System to latest software Solaris10(TM), Java ES(TM) 4 and Grid, Infrastructure Software along with seven paper publications. In addition, Lei has eight patents filed to US patent office and numerous patents pending at Sun. Lei has been actively contributing on Solaris10(TM), Grid, and SOA initiatives. Lei is a JCP member and a member of expert group for JSR 229, 235, 247, etc. Lei is a member of Enterprise Grid Alliance.

Lei's current research and development interests: Quantum Computer, Adaptive Computing, Algorithm and Information Theory, Operating System and Performance Model, Distributed Data Structure, Automata, Ubiquitous Computing, Mobile Ad-Hoc Network, Next Generation Internet, Parallel and Grid Computing, Virtualization, Attack and Detection, and Adaptive Middleware.

Reference

1. Canetti, et al. (2006). Time-bounded Task-PIOAs: A Framework for Analyzing Security Protocols. In the Proceedings of 20th International Symposium on Distributed Computing
2. Cantrill, B.M., Shapiro, M.W., Leventhal, A.H. (2004). In the Proceedings of USENIX Annual Technical Conference
3. Ishizu, et al. (2006). Adaptive Wireless-network Testbed for Cognitive Radio Technology. In the Proceedings of the 1st international workshop on Wireless Network Testbeds, Experimental Evaluation & Characterization.
4. Law, A.M., Kelton, W.D. (2000). Simulation Modeling and Techniques. 3rd ed. McGraw-Hill, New York.
5. Liu, X., Chien, A.A. (2003). Traffic-based Load Balance for Scalable Network Emulation. In the Proceedings of the ACM/IEEE Conference on Supercomputing.
6. Lundgren, et al. (2002). A Large-scale Testbed for Reproducible Ad hoc Protocol Evaluations. In the Proceedings of Wireless Communications and Networking Conference.
7. Shannon, C.E., McCarthy, J. (1956). Automata Studies. Princeton University Press, NJ.
8. Stonebrake, M. (1981). Operating System Support for Database Management. Communications of the ACM, 24(7), 412–418.

Performance Analysis of Interference for OFDM Systems

Jun Luo, Jean H. Andrian, Chi Zhou, and James P. Stephens, Sr.

Abstract The bit error rate (BER) analysis of various interference types is discussed for orthogonal frequency-division multiplexing (OFDM) systems in both analytical form and software simulation results. Specifically, the BER performance of barrage noise interference (BNI), partial band interference (PBI), and multitone interference (MTI) has been investigated in time-correlated Rayleigh fading channel with additive white Gaussian noise (AWGN). In addition, two novel intentional interference injecting methods – optimal-fraction PBI and optimal-fraction MTI – for OFDM systems are proposed with detailed theoretical analysis. Simulation results validate the analytical results. It is shown that under the various channel conditions, the optimal-fraction MTI always gives the best interference effect among all the interference models given in this chapter. Both analysis and simulation indicate that the proposed optimal-fraction MTI can be used to obtain improved interference effect under various channel conditions with low complexity for OFDM systems.

1 Introduction

Orthogonal frequency-division multiplexing (OFDM) is a promising technology that enables the transmission of high data rate. The basic idea of OFDM is to use a large number of parallel narrow-band sub-carriers instead of a single wide-band carrier to transport information. With its capability of adapting to severe channel conditions without complex equalization, OFDM is robust against inter-symbol interference (ISI) and fading caused by multipath propagation.

Since OFDM is a very promising candidate for the core technique of next-generation wireless communication systems, it is necessary to evaluate its

J. Luo (✉)
Department of Electrical and Computer Engineering, Florida International University, Miami, FL 33174, USA
e-mail:

S. Powell, J.P. Shim (eds.), *Wireless Technology*, Lecture Notes in Electrical Engineering 44, DOI 10.1007/978-0-387-71787-6_10,

performance under interference over fading channels. The interference can be unintentional, such as co-channel interference, or intentional, which is deliberately injected to disrupt the opponent's communications. This chapter focuses on the performance analysis of OFDM under the various interference models. There are some works done in this area. In [1], the performance of OFDM communication in the presence of partial-band interference is presented. The anti-interference property of clustered OFDM has been investigated in [2] and [3] gives a detailed study about the effect of partial band interference on OFDM systems. Despite of all the works mentioned, a comprehensive study about the effects of different interference models for OFDM systems has not been carried out. By comparing the bit error rate (BER) performance of different interference models, the most effective interference model can be identified under various channel conditions. This is very critical for both intentional interference injecting and anti-interference applications for OFDM systems.

This chapter evaluates the BER performance of different interference models, including Barrage Noise Interference (BNI), Partial Band Interference (PBI), and Multitone Interference (MTI) in time-correlated Rayleigh fading channel with Additive White Gaussian Noise (AWGN). In addition, two novel interference injecting methods – optimal-fraction PBI and optimal-fraction MTI – for OFDM systems are proposed with detailed theoretical analysis. The theoretical and simulation results show that the most effective interference injecting strategy for OFDM system is the optimal-fraction MTI since it can make interference effect better obviously through a simple way.

The chapter is organized as follows. In Section 2, a brief overview of OFDM system model is presented. Section 3 details the different interference models and their analytical BER forms in OFDM systems. Simulation results and related analysis are shown in Section 4. Finally, the concluding remarks are given in Section 5.

2 OFDM System and Channel Model

The overview of OFDM system model is illustrated in Fig. 1. In this chapter, we use binary phase shift keying (BPSK) and differential binary phase shift keying (DBPSK) as our signal mapping methods. Higher–order modulation techniques can be used as well, but these two are sufficient for us to analyze the essence of the problem. When applying DBPSK to OFDM systems, frequency domain differential demodulation (FDDD) is used rather than time domain differential demodulation (TDDD) since FDDD outperforms TDDD in frequency-nonselective fading channel [4]. The cyclic prefix is used as guard interval to eliminate ISI between the data blocks since samples of the channel output affected by this ISI can be discarded without any loss relative to the original information sequence [5].

The channel is modeled as a flat-fading Rayleigh channel. For every sub-carrier in OFDM systems, its bandwidth is relatively small compared with the bandwidth of the channel, so it is reasonable to make this flat-fading assumption. Based on

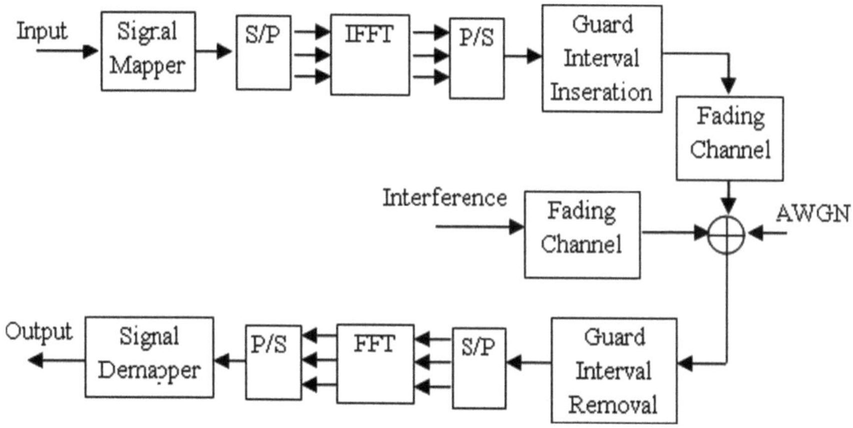

Fig. 1 OFDM system model

the method in [6], we construct a time-correlated flat-fading Rayleigh model, in which the channel has a Rayleigh-distributed envelope and uniform phase, and the two are mutually independent. We assume the system is coherent, so the phase can always be estimated perfectly. Hence we neglect the phase variation of the channel. Considering the interference and the signal are both independently attenuated by the channel, we use two independent random variables to describe the interference channel power gain G_J and the signal channel power gain G_s, which are given as

$$G_S = \alpha^2 \tag{1}$$

$$G_J = \beta^2 \tag{2}$$

where α and β are independent Rayleigh random variables with variances σ_S^2 and σ_J^2, respectively. Because the interference and signal are under the same channel environment, σ_S and σ_J can be regarded as the same value σ.

3 The Effects of Various Interference for OFDM Systems

In this section we investigate several typical interference models. For every interference type, we first give the BER form under AWGN, then consider more complicated Rayleigh fading channel.

3.1 BNI Under AWGN

Barrage noise interference (BNI) belongs to a broadband noise interference type. In this case, the interference interferes with the whole bandwidth by injecting a band-limited noise to the system. Its effect is the same as that of the AWGN noise, so the power spectrum density (PSD) of total noise becomes

$$PSD_N = N_0 + N_J \tag{3}$$

where N_0 is the noise PSD of complex AWGN noise and N_J is the PSD of complex BNI. Since the OFDM system performs no differently from conventional serial systems under the AWGN [7], the BER for BPSK and DBPSK is given as

$$P_{BPSK} = Q\left(\sqrt{\frac{2E_b}{N_0 + N_J}}\right) \tag{4}$$

$$P_{DBPSK} = \frac{1}{2}\exp\left(\frac{-E_b}{N_0 + N_J}\right) \tag{5}$$

where E_b is the average energy per bit of OFDM signal.

3.2 BNI Under Rayleigh Fading Channel with AWGN

In Rayleigh fading channel, the effective energy-per-bit becomes $G_S E_b$ and the effective PSD of the BNI becomes $G_J N_J$. Under the assumption that the time correlation coefficient of the channel is close to 1, the time-varying property of the channel will not affect the analysis of the differential modulation. After simplifying, we get the BER

$$P_{BPSK}(\alpha,\beta) = Q\left(\sqrt{\frac{2\alpha^2 E_b}{N_0 + \beta^2 N_J}}\right) \tag{6}$$

$$P_{DBPSK}(\alpha,\beta) = \frac{1}{2}\exp\left(\frac{-\alpha^2 E_b}{N_0 + \beta^2 N_J}\right) \tag{7}$$

Since there are two Rayleigh random variables with the same variance σ^2, the average BER for BPSK and DPBSK can be expressed as

$$\overline{P_{BPSK}} = \int_0^\infty \int_0^\infty Q\left(\sqrt{\frac{2\alpha^2 E_b}{N_0 + \beta^2 N_J}}\right) \cdot \frac{\alpha\beta}{\sigma^4} \cdot \exp\left(\frac{-\alpha^2 - \beta^2}{2\sigma^2}\right) d\alpha d\beta \tag{8}$$

$$\overline{P_{DBPSK}} = \int_0^\infty \int_0^\infty \frac{1}{2}\exp\left(\frac{-\alpha^2 E_b}{N_0 + \beta^2 N_J}\right) \cdot \frac{\alpha\beta}{\sigma^4} \cdot \exp\left(\frac{-\alpha^2 - \beta^2}{2\sigma^2}\right) d\alpha d\beta \tag{9}$$

Certainly, the infinite upper limit of integration should be replaced by finite approximated value in numerical calculation.

3.3 PBI Under AWGN

Partial band interference (PBI) is modeled as additive Gaussian noise with its power focusing on a portion of the entire bandwidth of the system. This strategy is considered more effective than BNI since the interference can use more power to interfere with certain specific bandwidth. We consider the best interference scenario: The interference signal bandwidth falls into that of the OFDM signal completely. The portion of interference signal bandwidth can be described by [1]

$$\rho = \frac{W_j}{W_{sig}} \tag{10}$$

where W_j is the bandwidth of the interference signal and W_{sig} is the bandwidth of the OFDM signal. To calculate the BER, we consider two types of frequency bands: the interfered frequency bands and the uninterfered frequency bands. Given the average PSD of PBI N_J, the effective PSD of PBI in the first type of bands becomes N_J/ρ, and there is no interference at all in the second type of bands. Combing those two cases with (4) and (5), the BER for BPSK and DBPSK under PBI is given as

$$P_{BPSK}(\rho) = \rho \cdot Q\left(\sqrt{\frac{2E_b}{N_0 + N_J/\rho}}\right) + (1-\rho)\cdot Q\left(\sqrt{\frac{2E_b}{N_0}}\right) \tag{11}$$

$$P_{DBPSK}(\rho) = \frac{\rho}{2}\cdot \exp\left(\frac{-E_b}{N_0 + N_J/\rho}\right) + \left(\frac{1-\rho}{2}\right)\cdot \exp\left(\frac{-E_b}{N_0}\right) \tag{12}$$

Since (11) and (12) depend on the value of ρ, it is necessary to find the optimal interference fraction ρ^* so that the interference effect is maximized. Assuming the background AWGN is negligible compared to PBI and applying the following approximation for the Q function from [8]:

$$Q(x) \approx \exp(-\frac{x^2}{2})/(1.64x + \sqrt{0.76x^2+4}) \tag{13}$$

Equations (11) and (12) are transformed into

$$\hat{P}_{BPSK}(\rho) = \rho \frac{\exp(-SIR\cdot\rho)}{(1.64\sqrt{2SIR\cdot\rho} + \sqrt{1.52\cdot SIR\cdot\rho + 4})} \tag{14}$$

$$\hat{P}_{DBPSK}(\rho) = \frac{\rho}{2}\exp(-SIR\cdot\rho) \tag{15}$$

where SIR represents signal-to-interference ratio, which is equal to E_b/N_J. The optimal interference fraction ρ^* can be obtained by maximizing (14) and (15) with respect to ρ for a given *SIR*

$$\rho^*_{BPSK} = \underset{\rho}{\operatorname{argmax}}\, \hat{P}_{BPSK}(\rho) \tag{16}$$

$$\rho^*_{DBPSK} = \underset{\rho}{\operatorname{argmax}}\, \hat{P}_{DBPSK}(\rho) \tag{17}$$

(interference fraction constraint: $0 \leq \rho \leq 1$)

Let us take the partial derivative of (14) and (15) with respect to ρ and set them to 0 to obtain optimal interference fractions. Here Newton–Raphson approximation is used to get numerical results. Like constrained control system, the optimal interference fraction will saturate whenever the boundary constraints ($0 \leq \rho \leq 1$) are violated. In practice, the optimal values of interference fraction can be generated offline based on different *SIR*, and then stored in hardware or software as a table. By looking up this table, the interference generator can maintain optimal performance for every value of the *SIR*. In this chapter, this special PBI based on optimal interference fraction table is named as optimal-fraction PBI.

3.4 PBI Under Rayleigh Fading Channel with AWGN

For the time-correlated Rayleigh-fading channel, following the same steps as before, the average BER for BPSK and DPBSK becomes

$$\begin{aligned}\overline{P_{BPSK}} = &\int_0^\infty\int_0^\infty \left(\rho \cdot Q\left(\sqrt{\frac{2\alpha^2 E_b}{N_0 + \beta^2 N_J/\rho}}\right) + (1-\rho)\cdot Q\left(\sqrt{\frac{2\alpha^2 E_b}{N_0}}\right)\right) \\ &\times \tfrac{\alpha\beta}{\sigma^4}\cdot \exp\left(\tfrac{-\alpha^2-\beta^2}{2\sigma^2}\right) d\alpha d\beta\end{aligned} \tag{18}$$

$$\begin{aligned}\overline{P_{DBPSK}} = &\int_0^\infty\int_0^\infty \left(\frac{\rho}{2}\cdot \exp\left(\frac{-\alpha^2 E_b}{N_0 + \beta^2 N_J/\rho}\right) + \left(\frac{1-\rho}{2}\right)\cdot \exp\left(\frac{-\alpha^2 E_b}{N_0}\right)\right) \\ &\times \tfrac{\alpha\beta}{\sigma^4}\cdot \exp\left(\tfrac{-\alpha^2-\beta^2}{2\sigma^2}\right) d\alpha d\beta\end{aligned} \tag{19}$$

Here we do not consider the optimization of interference fraction, since this requires a complicated algorithm in a time-varying Rayleigh fading channel. It is not realistic to do so just for small improvement in interference effect. Instead we introduce an empirical equation to get the approximation solution. As Rayleigh fading channel is a special form of Rician fading channel, the commonly used Rician fading amplitude with unit mean-squared value is defined as [9]

$$\gamma_i = \sqrt{\frac{(x_i + \sqrt{2\,K})^2 + y_i^2}{2(K+1)}} \tag{20}$$

where x_i and y_i are samples of zero-mean stationary Gaussian random processes with variance $\sigma_0^2 = 1$, which can be generated by the method in [6]. The best and worst-case Rician fading channels associated with K-factors of $K = \infty$ and $K = 0$ are the Gaussian and Rayleigh channels with strong line of sight (LOS) and without LOS path, respectively. From the previous section, it is easy to get the optimal interference fractions in *AWGN channels:* $\rho^*_{BPSK\_AWGN}$ and $\rho^*_{DBPSK\_AWGN}$ (unconstrained values from Eqs. (16) and (17)). Based on these values, we can get the optimal interference fractions in Rician channel by empirical fit as

$$\rho^*_{BPSK\_RICIAN} = (1 + e^{-0.0123\,K^2 - 0.148\,K + 2.20})\rho^*_{BPSK\_AWGN} \tag{21}$$

$$\rho^*_{DBPSK\_RICIAN} = (1 + e^{-0.0023\,K^2 - 0.2543\,K + 2.08})\rho^*_{DBPSK\_AWGN} \tag{22}$$

(Interference fraction constraint:$0 \leq \rho \leq 1$)

3.5 *MTI Under AWGN*

Multitone interference (MTI) divides its total power into q distinct, equal power, random phase tones. Every interference tone can be modeled as

$$J(t) = A_J e^{j(2\pi f_J t + \phi_J)} \tag{23}$$

where ϕ_J is the random phase, which is uniformly distributed over [0,2π]. A_J and f_J are the amplitude and frequency, respectively. We assume that those q interference tones are perfectly aligned with q sub-carriers of the OFDM system. Then the portion of interference signal bandwidth is defined as

$$\rho = q/M \tag{24}$$

where M is the number of FFT points. After FFT in receiver block, in signal space, the interference signal, which has fixed length A_J and random phase ϕ_J, is added to original signal (S_1 or S_2) as a vector (Fig. 2), and AWGN noise $\vec{V}$ is another vector added to them. Projecting the compound signal onto ϕ_1 axis, we get

$$S_{\phi 1,r1} = A_J \cos(\phi_J) + \sqrt{E_b} + V_{\phi 1} \tag{25}$$

$$S_{\phi 1,r2} = A_J \cos(\phi_J) - \sqrt{E_b} + V_{\phi 1} \tag{26}$$

where $S_{\phi 1,r1}$ and $S_{\phi 1,r2}$ represent the ϕ_1 axis projection of received signals (corresponding to S_1 and S_2, respectively), and $V_{\phi 1}$ is the AWGN noise $\vec{V}$'s ϕ_1 axis projection. Because the error probabilities for S_1 and S_2 are equal, only $P_{e_{s1}}$ needs to be calculated. Hence the BER for BPSK with MTI is given as

$$P_{BPSK\_MTJ} = P_{e_{s1}} = \Pr\ (S_{\phi 1,r1} < 0) \tag{27}$$

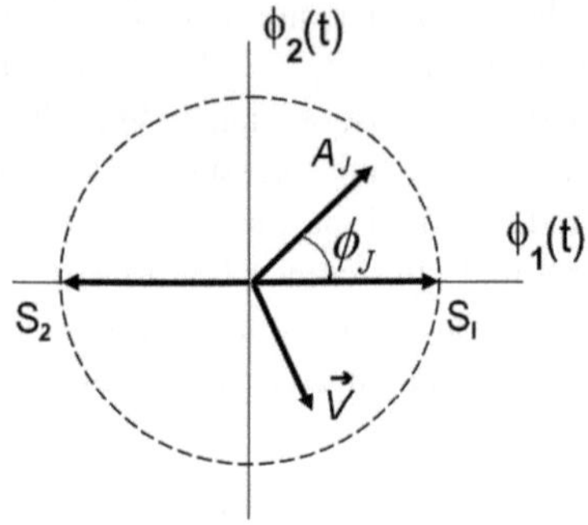

Fig. 2 OFDM system model

Here we have two random variables: one is ϕ_J, which is uniformly distributed over $[0,2\pi]$ and the other is $V_{\phi 1}$, which satisfies the Gaussian distribution with $N(0, N_0/2)$. Now we define $Y = A_J \cos(\phi_J) + \sqrt{E_b}$, $X = V_{\phi 1}$ and $W = X + Y$. Then

$$P_{BPSK\_MTJ} = \Pr\ (S_{\phi 1, r1} < 0) = \Pr\ (W < 0) \tag{28}$$

The probability density function (PDF) of $cos(\theta)$ function is given in [10] as

$$f_Z(z) = \begin{cases} \frac{1}{\pi} \frac{1}{\sqrt{1-z^2}} & z \in (-1,1) \\ 0 & \end{cases} \tag{29}$$

The PDF of Y can be represented as

$$f_Y(y) = \frac{1}{\pi A_J \sqrt{1 - (\frac{y-\sqrt{E_b}}{A_J})^2}} \tag{30}$$

where y should satisfy $\sqrt{E_b} - A_J < y < \sqrt{E_b} + A_J$

Given the PDF of X, the PDF of W is

$$\begin{aligned} f_W(w) &= \int_{-\infty}^{+\infty} f_X(w-y) f_Y(y) dy \\ &= \int_{\sqrt{E_b}-A_J}^{\sqrt{E_b}+A_J} \frac{1}{\sqrt{\pi N_0}} \exp\left(-\frac{(w-y)^2}{N_0}\right) \frac{1}{\pi A_J \sqrt{1 - (\frac{y-\sqrt{E_b}}{A_J})^2}} dy \end{aligned} \tag{31}$$

Then A_J is the only unknown variable in (31), which can be obtained simply through the following equation

$$A_J = \sqrt{\frac{E_b}{\rho SIR}} \tag{32}$$

Based on the result of (31), we can get P_{BPSK_MTJ} from (28) easily. Thus the BER for BPSK is given as

$$\begin{aligned} P_{BPSK}(E_b,N_0,\rho,SIR) &= \rho \cdot P_{BPSK\_MTJ} + (1-\rho)Q\left(\sqrt{\frac{2E_b}{N_0}}\right) \\ &= \rho \int_{-\infty}^{0} f_W(w)dw + (1-\rho)Q\left(\sqrt{\frac{2E_b}{N_0}}\right) \end{aligned} \tag{33}$$

If the AWGN noise is negligible, we can get

$$P_{BPSK}(E_b,\rho,SIR) = P_{es1} = \begin{cases} \frac{\rho}{\pi}(\frac{\pi}{2} - \arcsin(\sqrt{\rho SIR})) & \sqrt{\rho SIR} < 1 \\ 0 & \end{cases} \tag{34}$$

For DBPSK, let $z(k-1)$ and $z(k)$ be the reference and received symbol vectors, respectively. The variable D is given as

$$\begin{aligned} &D = \mathrm{Re}(z(k)\, z^*(k-1)) = \\ &Re((A_J e^{j\phi_J(k)} + s(k) + \vec{V}(k))(A_J e^{-j\phi_J(k-1)} + s^*(k-1) + \vec{V^*}(k-1))) \end{aligned} \tag{35}$$

where $s(k)$and $s(k-1)$ are transmitted constellations at time k and $k-1$. D is used as differential detector to decide which symbol was transmitted. Due to symmetry, we can assume a given phase difference zero to compute the error probability [5], so $s(k)$and $s(k-1)$ can be specified as$\sqrt{E_b}$. Hence

$$\begin{aligned} &D = \mathrm{Re}(z(k)\, z^*(k-1)) = \\ &\mathrm{Re}((A_J e^{j\phi_J(k)} + \sqrt{E_b} + \vec{V}(k))(A_J e^{-j\phi_J(k-1)} + \sqrt{E_b} + \vec{V^*}(k-1))) \end{aligned} \tag{36}$$

If D is less than 0, then a decision error is made. That is, the BER equals to $\Pr(D < 0)$ when MTI exists. However, the probability of D is difficult to calculate from (36), subsequently some approximations are necessary. For $SIR >> SNR$ (E_b/N_0), we can neglect some small items in (36) and get

$$\begin{aligned} D \approx E_b &+ A_J\sqrt{E_b}(\cos(\phi_J(k)) + \cos(\phi_J(k-1))) + \sqrt{E_b}(V_{\phi 1}(k) + V_{\phi 1}(k+1)) \\ &+ A_J^2(\cos(\phi_J(k) - \phi_J(k-1))) \end{aligned} \tag{37}$$

Dividing (37) by $\sqrt{E_b}$ yields

$$\begin{aligned}\overline{D} \approx & \sqrt{E_b} + A_J(\cos(\phi_J(k)) + \cos(\phi_J(k-1))) + V_{\phi 1}(k) + V_{\phi 1}(k-1) \\ & + \frac{A_J^2}{\sqrt{E_b}}(\cos(\phi_J(k) - \phi_J(k-1)))\end{aligned} \tag{38}$$

Compared with the case of coherent BPSK (25), there are five noise terms instead of two. Approximately, the MTI is $2+1/\sqrt{\rho SIR}$ times larger than that of BPSK and AWGN is two times larger than that of BPSK, which gives a simple way of getting differential modulation BER from coherent modulation BER in Eq. (33):

$$P_{DBPSK}(E_b,N_0,\rho,SIR) = P_{BPSK}\left(E_b,2N_0,\rho,\frac{SIR}{(2+\frac{1}{\sqrt{\rho SIR}})}\right) \tag{39}$$

This equation is valid only for *SIR* >> *SNR*. From simulation, we found that when *SIR* is close to *SNR*, the simulation values of BER will deviate from theoretical values to some small values. In order to compensate this deviation, Eq. (39) was modified empirically to

$$P_{DBPSK}(E_b,N_0,\rho,SIR) = P_{BPSK}\left(E_b,N_0,\rho,\frac{SIR}{(2+\frac{1}{\sqrt{\rho SIR}}-0.6)}\right) \tag{40}$$

which shows good results in the simulation.

In MTI, optimal interference fraction should also be considered, so that we get the optimal-fraction MTI. Again, assuming the background AWGN is negligible, from (34) following the same process as PBI, we can get

$$\frac{1}{\pi}\left(\frac{\pi}{2} - \arcsin\left(\sqrt{\rho^*_{BPSK}\cdot SIR}\right)\right) + \frac{\rho^*_{BPSK}}{\pi}\frac{-0.5\cdot SIR}{\sqrt{\rho^*_{BPSK}\cdot SIR\,(1-\rho^*_{BPSK}\cdot SIR)}} = 0 \tag{41}$$

Solving it with different SIR will generate MTI interference fraction optimization table of BPSK. For DBPSK, the optimal interference fraction can be obtained from simulation results. The optimization tables of BPSK and DBPSK will be used by optimal-fraction multitone interference generator to achieve optimal performance.

3.6 MTI Under Rayleigh Fading Channel with AWGN

Similar as before, the average BER for BPSK and DBPSK for MTI under Rayleigh fading channel with AWGN noise is

$$\overline{P_{BPSK}} = \int_0^\infty\int_0^\infty (P_{BPSK}(\alpha^2 E_b,N_0,\rho,\alpha^2 SIR/\beta^2))\frac{\alpha\beta}{\sigma^4}\cdot\exp\left(\frac{-\alpha^2-\beta^2}{2\sigma^2}\right)d\alpha d\beta \tag{42}$$

$$\overline{P_{DBPSK}} = \int_0^{\infty}\int_0^{\infty} (P_{DBPSK}(\alpha^2 E_b,N_0,\rho,\alpha^2 SIR)\big/\beta^2)\frac{\alpha\beta}{\sigma^4}\cdot\exp\left(\frac{-\alpha^2-\beta^2}{2\sigma^2}\right)d\alpha d\beta \tag{43}$$

where $P_{BPSK}(\alpha^2 E_b,N_0,\rho,\alpha^2 SIR\big/\beta^2)$ is derived from (33), and when you calculate it, you should change all E_b and SIR to $\alpha^2 E_b$, $\alpha^2 SIR\big/\beta^2$ correspondingly. Equation (40) gives the value of $P_{DBPSK}(\alpha^2 E_b,N_0,\rho,\alpha^2 SIR\big/\beta^2)$. Here E_b and SIR should be changed as well. Also, given the MTI optimal interference fractions in AWGN channels: $\rho^*_{BPSK\_AWGN}$ and $\rho^*_{DBPSK\_AWGN}$ (unconstrained values), the optimal interference fractions in Rician channel can be got empirically as

$$\rho^*_{BPSK\_RICIAN} = (1 + e^{-0.0187\,K^2 - 0.3528\,K + 2.4})\rho^*_{BPSK\_AWGN} \tag{44}$$

$$\rho^*_{DBPSK\_RICIAN} = (1 + e^{-0.0131\,K^2 - 0.308\,K + 2.49})\rho^*_{DBPSK\_AWGN} \tag{45}$$

$$(\text{interference fraction constraint:} 0 \le \rho \le 1)$$

4 Simulation Results and Analysis

In this section, the BER performance of different interference techniques for OFDM system is evaluated by the means of software simulation. Based on the 802.11a standard [11], the main parameters used in the simulation are summarized as Table 1. In this table, to simplify the problem, we use 64 as the number of sub-carriers instead of 52 in the 802.11a standard. Hence the occupied bandwidth is changed from 16.6 to 20 MHz correspondingly.

Figure 3 shows the comparison between simulation results and theoretical results of all non-optimal interference types (BNI, the fixed-fraction PBI and the fixed-fraction MTI) in the chapter. In the simulation, every test is repeated 200–1000 times to eliminate the fluctuations caused by intrinsic random nature of the OFDM

Table 1 Main parameters used in simulations

Signal bit rate	20 MHz	Modulation scheme	BPSK/DBPSK
Number of sub-carriers	64	Cyclic prefix	0.8 us
FFT length	64	Channel model	Rayleigh fading channel with AWGN noise
Doppler frequency	40 Hz	OFDM symbol period	3.2 us
Signal bandwidth	20 MHz	Interference bandwidth	Depend on different interference techniques

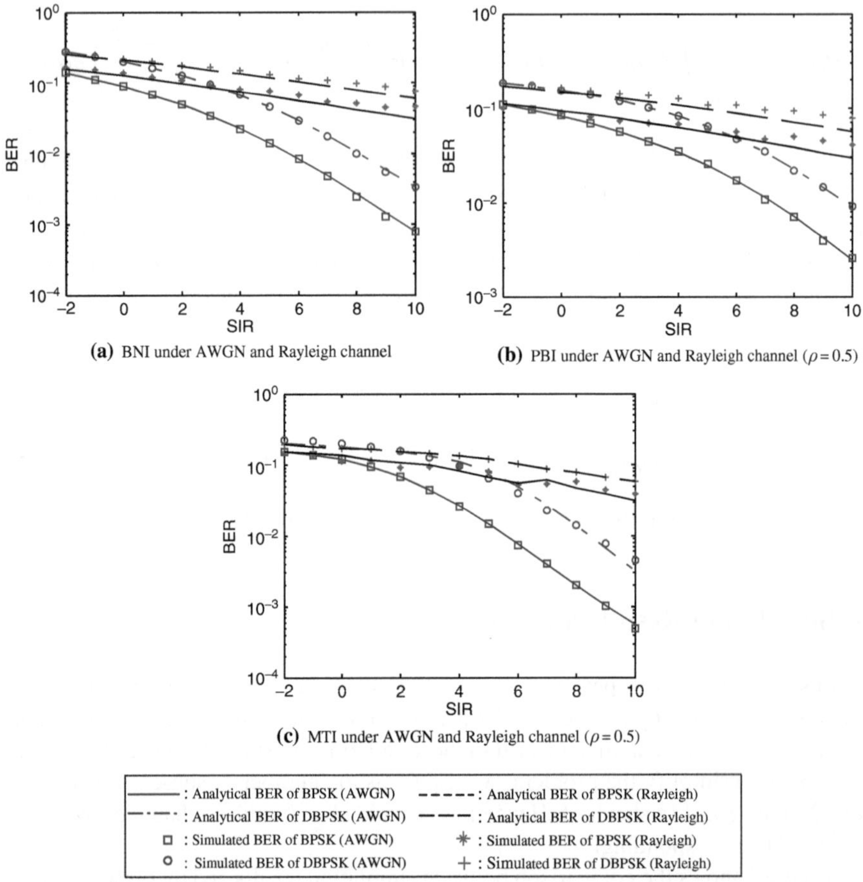

(a) BNI under AWGN and Rayleigh channel

(b) PBI under AWGN and Rayleigh channel ($\rho = 0.5$)

(c) MTI under AWGN and Rayleigh channel ($\rho = 0.5$)

Fig. 3 Comparison between simulation results and theoretical values of all interference strategies (*SNR* is fixed to 10 dB)

communication system. It is shown that the simulation results of all non-optimal interference types are in agreement with the analytical prediction perfectly under AWGN. On the contrary, in Rayleigh fading channel, there are some small deviations between simulation results and theoretical values. They are caused by precision errors of numerical integration (for all non-optimal interference types) and equation approximation (only for DBPSK of MTI). All these deviations are less than 10%, therefore they are acceptable.

To verify the optimization process about PBI and MTI, optimal interference fraction data are listed as Table 2, in which *SNR* is fixed to 20 db and *SIR* is varied from –2 to 10 db. We show both analytical predictions and simulated values for PBI and MTI except MTI for DBPSK, for which only simulated values are shown since precise theoretical equation is hard to obtain. Since 64-FFT is used in the proposed OFDM system, every ρ^* has been rounded to the closest integer multiple of 1/64.

Table 2 Optimal interference fraction ($SNR = 20$ dB)

SIR [dB]	ρ^* PBI BPSK (A)	ρ^* PBI BPSK (S)	ρ^* PBI DBPSK (A)	ρ^* PBI DBPSK (S)	ρ^* MTI BPSK (A)	ρ^* MTI BPSK (S)	ρ^* MTI DBPSK (S)
−2	**1**	62/64	**1**	62/64	**1**	1	1
−1	**57/64**	57/64	**1**	1	**51/64**	48/64	63/64
0	**45/64**	46/64	**1**	61/64	**40/64**	41/64	55/64
1	**36/64**	35/64	**51/64**	52/64	**32/64**	32/64	43/64
2	**29/64**	31/64	**40/64**	41/64	**25/64**	24/64	32/64
3	**23/64**	26/64	**32/64**	32/64	**20/64**	20/64	27/64
4	**18/64**	21/64	**25/64**	25/64	**16/64**	15/64	21/64
5	**14/64**	15/64	**20/64**	21/64	**13/64**	12/64	16/64
6	**11/64**	11/64	**16/64**	17/64	**10/64**	11/64	12/64
7	**9/64**	9/64	**13/64**	14/64	**9/64**	8/64	10/64
8	**7/64**	7/64	**10/64**	10/64	**6/64**	7/64	8/64
9	**6/64**	6/64	**8/64**	8/64	**5/64**	5/64	6/64
10	**5/64**	4/64	**6/64**	7/64	**4/64**	4/64	5/64

A: analytical values, *S*: simulated values
This table is generated under AWGN channel without channel fading

In Table 2, the bold part is the analytical prediction and its right side is the corresponding simulated results. The biggest error between them is less than 5%, which validates the correctness of the analytical model.

The comparison between the optimal-fraction interference and the fixed-fraction interference of PBI and MTI is shown in Fig. 4. It is revealed that under AWGN and Rayleigh fading channel, the optimal-fraction interference always gives the best interference effect. Also it is found that the 0.9-fixed-fraction interference almost gives the best interference effect in Rayleigh fading channel. In fact, in deep fading channel, the interference power should be distributed to the whole bandwidth to gain the best interference effect, so the optimal-fraction interference degrades to whole bandwidth interference in this case. In general, the optimal-fraction interference gives us a simple way to obtain good interference effect under various channel conditions.

Finally, optimal-fraction MTI and optimal-fraction PBI are compared in Figs. 5 and 6. Under various channel conditions shown in the Figs. 5 and 6, the optimal-fraction MTI clearly outperforms the optimal-fraction PBI for BPSK and DBPSK. The advantage of the optimal-faction MTI is most obvious in AWGN channel. As the channel condition gets worse, the advantage dies down gradually (see the Rician channel with $K = 5$ and 10 in Fig. 6). In worst fading case – Rayleigh fading channel, the optimal-faction MTI still outperforms the optimal-fraction PBI with a small advantage. Hence it is reasonable to believe that the optimal-fraction MTI can be used to obtain improved interference effect under different channel conditions with low complexity.

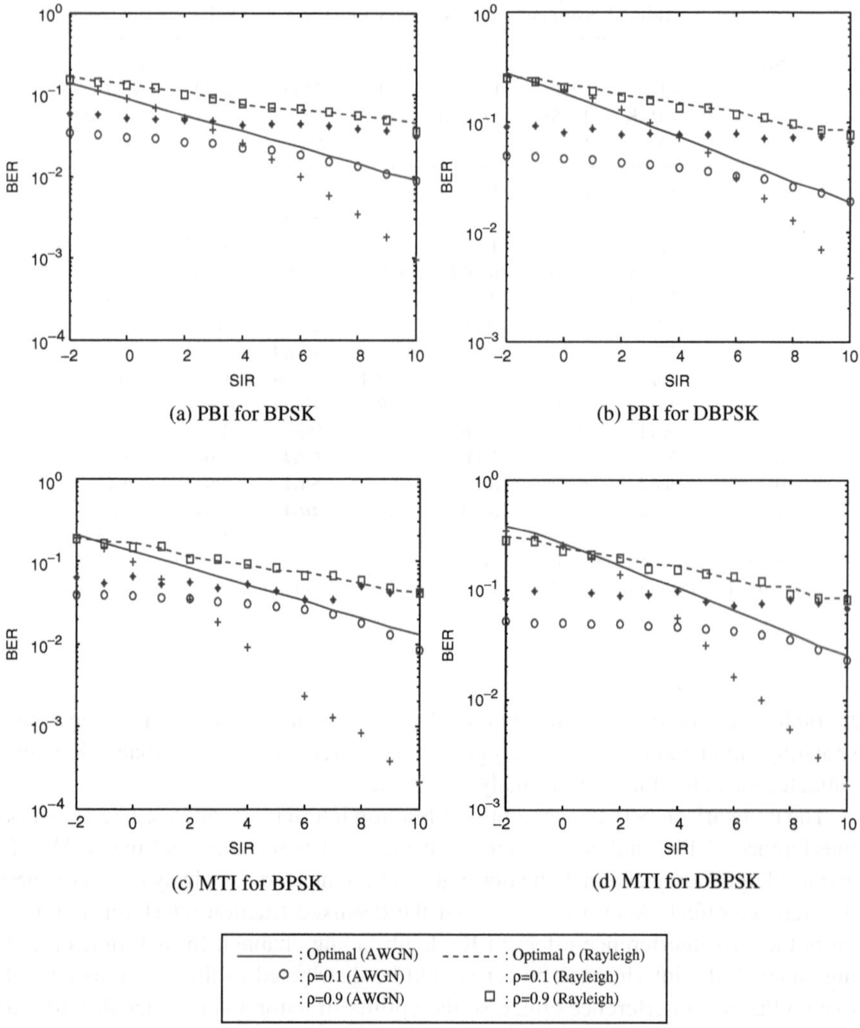

(a) PBI for BPSK (b) PBI for DBPSK (c) MTI for BPSK (d) MTI for DBPSK

Fig. 4 Comparison between optimal-fraction interference and fixed-fraction interference of PBI and MTI (*SNR* is fixed to 10 dB)

5 Conclusion

The BER performance of different interference strategies for OFDM system has been investigated. Both analytical form and simulation values are given. In addition, two new interference injecting methods – optimal-fraction PBI and optimal-fraction MTI are proposed in this chapter. Through analysis and simulation, it is shown that under the various channel conditions, the optimal-fraction MTI clearly outperforms other interference strategies in the chapter. The results of the experiment and the

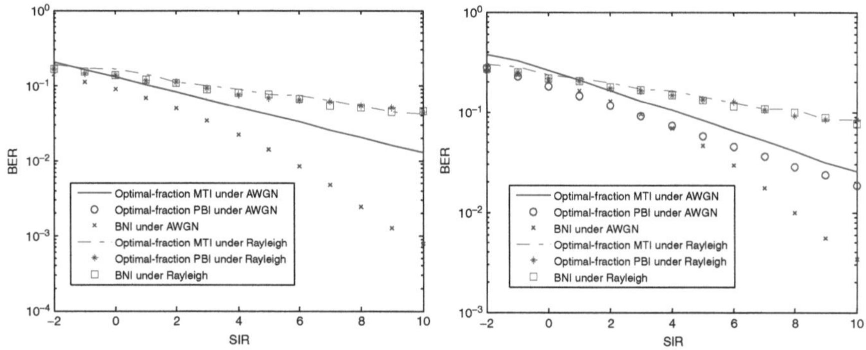

(a) Comparison for BPSK (b) Comparison for DBPSK

Fig. 5 Comparison of optimal-fraction MTI, optimal-fraction PBI, and BNI (*SNR* is fixed to 10 dB in AWGN and Rayleigh channel)

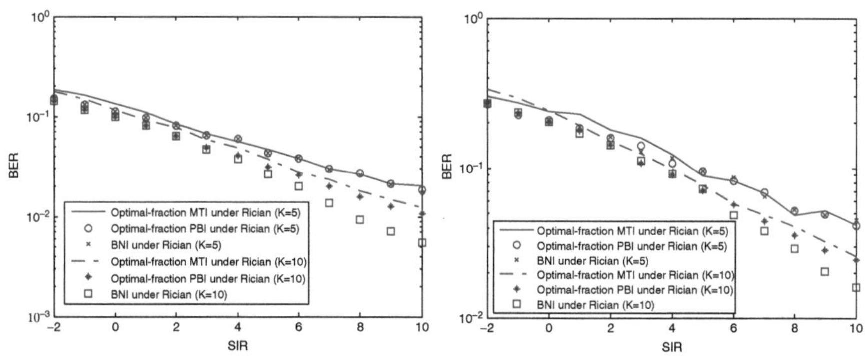

(a) Comparison for BPSK (b) Comparison for DBPSK

Fig. 6 Comparison of optimal-fraction MTI, optimal-fraction PBI, and BNI (*SNR* is fixed to 10 dB in Rician channel, $K = 5$ and 10)

analysis of those results show that the optimal-fraction MTI is a very effective interference injecting technique for OFDM system in various channel conditions.

References

1. R. F. Ormondroyd and E. Al-Susa, "Impact of multipath fading and partial-band interference on the performance of a COFDM/CDMA modulation scheme for robust wireless communications," IEEE MILCOM, 2, 673–678, 1998.
2. H. Zhang and Y. Li, "Anti-jamming property of clustered OFDM for dispersive channels," IEEE MILCOM, 1, 336–340, October 2003.
3. J. Park et al., "Effect of partial band jamming on OFDM-based WLAN in 802.11 g," ICASSP 2003, 4, 560–563, April 2003.
4. S. Lijun et al., "BER Performance of frequency domain differential demodulation OFDM in flat fading channel," GLOBECOM, 1, 1–5, 2003.

5. A. Goldsmith, *Wireless Communications*, Cambridge University Press, Cambridge, 2005.
6. Y.R. Zheng and C. Xiao, "Improved models for the generation of multiple uncorrelated Rayleigh fading waveforms," IEEE Communications Letters, 6, 6, 256–258, 2002.
7. L. Hanzo et al., *OFDM and MC-CDMA for Broadband Multi-User Communications, WLANs and Broadcasting*, Wiley-IEEE Press, US, September 2003.
8. N. Kingsbury, "Approximation Formulae for the Gaussian Error Integral, Q(x)," Connexions, June 7, 2005.
9. N. Kostov, "Mobile radio channels modeling in Matlab," Journal of Radioengineering, 12, part 4, 12–17, 2003.
10. A. Papoulis, *Probability, Random Variables and Stochastic Processes*, 3rd Edition, McGraw-Hill, New York, Feb. 1991
11. IEEE 802.11a, "Part 11: Wireless LAN Medium Access Control (MAC) and Physical Layer (PHY) Specifications: High-speed Physical Layer in the 5 GHz Band," supplement to IEEE 802.11 Standard, September 1999.

Maximum-Likelihood Carrier-Frequency Synchronization and Channel Estimation for MIMO-OFDM Systems

Soheil Salari, Mahmoud Ahmadian, Mehrdad Ardebilipour, Vahid Meghdadi, and Jean-Pierre Cances

Abstract In this chapter, we propose a new scheme for maximum-likelihood (ML) estimation of both carrier-frequency offset (CFO) and channel coefficients in multiple-input multiple-output (MIMO) orthogonal frequency division multiplexing (OFDM) systems, assuming that a training sequence is available. Our scheme is also capable to accommodate any space–time coded (STC)-OFDM transmission. Furthermore, the Cramer–Rao bounds (CRBs) for both CFO and channel estimators are developed to evaluate the performance of the proposed scheme. The simulation results show that the proposed algorithm achieves almost ideal performance compared with the CRB for both channel and frequency offset estimations.

1 Introduction

High data-rate wireless access is demanded by many applications. Traditionally, more bandwidth is required for higher data-rate transmission. However, due to spectral limitations, it is often impractical or sometimes very expensive to increase bandwidth. In this case, using multiple transmit and receive antennas for spectrally efficient transmission is an alternative solution. Multiple transmit antennas can be used either to obtain transmit diversity or to form multiple-input multiple-output (MIMO) channels.

Many researchers have studied using multiple transmit antennas for diversity in wireless systems. Transmit diversity may be based on linear transforms [1] or space–time coding [2]. In particular, space–time coding is characterized by high code efficiency and good performance; hence, it is a promising technique to improve the efficiency and performance of wireless communication systems. On the other hand, the system capacity can be significantly improved if multiple transmit and receive antennas are used to form MIMO channels [3–6]. It is proven in [4] that compared

S. Salari (✉)
K. N. Toosi University of Technology, Faculty of Electrical Engineering, Tehran, Iran.
e-mail: salari@eetd.kntu.ac.ir

S. Powell, J.P. Shim (eds.), *Wireless Technology*, Lecture Notes in Electrical Engineering 44, DOI 10.1007/978-0-387-71787-6_11,

with a single-input single-output (SISO) system with flat Rayleigh fading or narrowband channels, a MIMO system can improve the capacity by a factor of the minimum number of transmit and receive antennas.

The combination of MIMO signal processing with orthogonal frequency-division multiplexing (OFDM) has gained considerable interest in recent years [7, 8]. MIMO offers extraordinary throughput without additional power consumption or bandwidth expansion [9], and OFDM introduces overlapping but orthogonal narrowband subchannels to convert a frequency selective fading channel into a non-frequency selective one. Moreover, OFDM avoids inter-symbol interference (ISI) by means of cyclic prefix (CP) [10]. Hence, in the presence of frequency selectivity, it is beneficial to consider MIMO in the OFDM context [11].

It is known that similar to single antenna OFDM, MIMO-OFDM is very sensitive to the frequency synchronization and channel estimation errors [12]. Carrier-frequency offset (CFO) induced by the mismatches of local oscillators in transmitter and receiver causes inter-carrier interferences (ICI), which may result in significant performance degradation. Several carrier-frequency synchronization schemes for MIMO-OFDM systems are reported in the literature [13–15]. Moreover, the coherent detection of MIMO-OFDM signals requires channel estimation to mitigate amplitude and phase distortions in a fading channel. Various channel estimation algorithms are also proposed for MIMO-OFDM systems (see, e.g., [12, 16, 17] and references therein).

In dealing with channel estimation, most investigators assume zero frequency offset between the carrier and the local reference at the receiver. In practice, this means that the offset is so small that the demodulated signal incurs only negligible phase rotations. Using stable oscillators is not a viable route to meet such conditions for; in general, the stability requirements would be too stringent. Furthermore, even ideal oscillators would be inadequate in a mobile communication environment experiencing significant Doppler shifts. The only solution is to measure the CFO accurately [18]. The combination of CFO and channel estimation leads to particularly complex problems in MIMO-OFDM systems due to the number of unknowns [12].

More recently, joint channel and frequency offset estimation issue has received a lot of attentions [19, 20] in OFDM context. The exact maximum-likelihood (ML) solutions of both frequency offset and channel impulse response (CIR) are prohibitively complex. Therefore, in [19], the ML estimate for only frequency offset was obtained based on the least square (LS) CIR estimate. In [20], an adaptive approach (i.e., steepest descent algorithm) was employed to avoid the complexity of joint ML estimation, where first the channel estimation is performed assuming that the frequency offset is known, and then the frequency offset is estimated assuming the channel state is known. However, all of these estimators [19, 20] are designed for SISO-OFDM systems rather than MIMO systems. Recently, Pun et al. in [21] proposed a new method for ML synchronization and channel estimation in OFDMA uplink transmission based on the alternating projection algorithm [22].

In this chapter, we present a reduced-complexity scheme for ML estimate of both CFO and CIR in multi antenna OFDM transmission, assuming that a training sequence is available. As we shall see, the solution consists of two separate steps: a CFO estimator and a channel estimator. It is known that the expectation-maximization (EM) algorithm [23, 24] can provide the ML solutions in an iterative manner for ML estimation problems [25, 26]. Therefore, to overcome the difficulty of ML estimation of CFO, we resort to the EM algorithm and propose a novel EM-based CFO estimator (first step). The CFO estimates are then exploited in the second step to estimate the MIMO channel coefficients. Our scheme is also capable to accommodate any space-time coded (STC) transmission. Moreover, to benchmark the performance of the proposed scheme, the Cramer–Rao bounds (CRBs) are derived for both CFO and CIR estimators. Simulation results show that the proposed algorithm achieves almost ideal performance compared with the CRBs in all ranges of signal-to-noise ratio (SNR) for both channel and frequency offset estimates. The need for a training block at the beginning of each frame leads to an increased system overhead. However, this should not be a serious concern since training blocks are specified in the frame structure of many standardized multicarrier systems [21].

The rest of the chapter is organized as follows. In the next section, we describe the MIMO-OFDM signal model with the presence of CFO. Section 3 is devoted to the description of the ML estimate of both CFO and CIR. The performance of the estimators is investigated in Section 4. Simulation results are given in Section 5 and conclusion summarizes the main results in Section 6.

2 MIMO-OFDM Signal Model in the Presence of CFO

We consider a MIMO-coded OFDM communication system with K subcarriers, N transmit (Tx), and M receive (Rx) antennas, signaling through an L-tap frequency-selective fading channel in the presence of frequency offset. As illustrated in Fig. 1, the information bits are first encoded by encoder into coded bits and then the coded

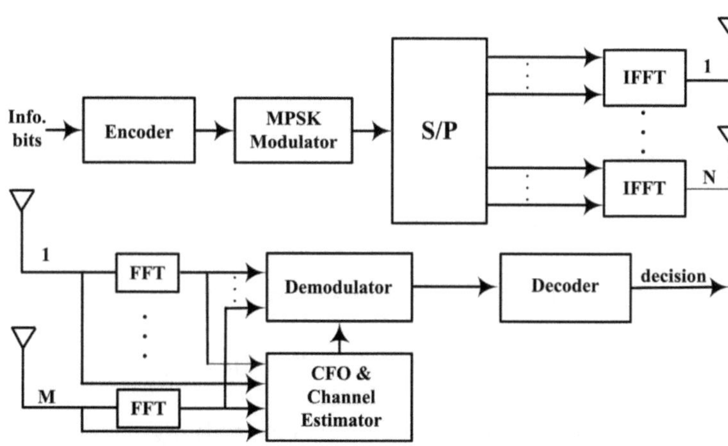

Fig. 1 MIMO-OFDM system model

bits are modulated into MIMO symbols by an MPSK modulator. These MIMO symbols are split into N sequences of K symbols; the K MIMO symbols of each sequence are transmitted from one particular transmit antenna at K subcarriers over one OFDM slot.

As in a typical data communication scenario, communication is carried out in a burst manner. A data burst structure is illustrated in Fig. 2. It spans $P+1$ OFDM blocks, with the first OFDM block containing known training symbols. The remaining P OFDM blocks contain data symbols. Note that each OFDM block consists of NK MIMO symbols, which are simultaneously transmitted during one time slot.

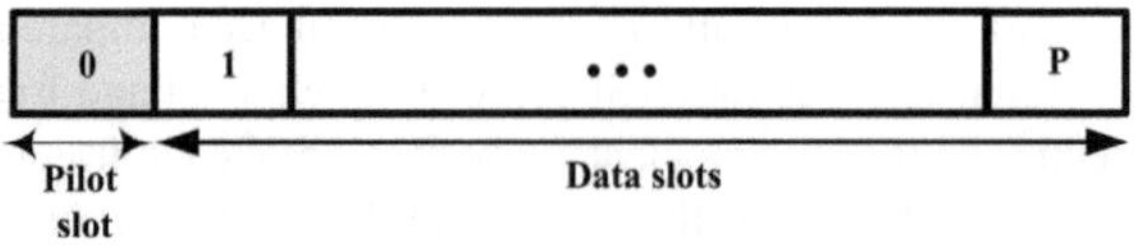

Fig. 2 Frame data burst structure

It is assumed that the fading process is static during each OFDM block (one time slot) but it varies from one OFDM block to another, and the fading processes associated with different transmit – receive antenna pairs are uncorrelated. The Lth order frequency-selective fading channel from the jth transmit antenna to the ith receive antenna at the pth time slot is denoted by [21]

$$h_{i,j}[p] = \left[h_{i,j}[p,0], \ldots, h_{i,j}[p,L-1]\right]^T_{L\times 1}, 1 \le j \le M, 1 \le i \le N \tag{1}$$

At each receiving antenna, a superposition of faded signals from all transmit antennas plus noise is received. Assuming ideal synchronization in time, the received signal at ith receive antenna can be expressed as [15]

$$y_i[p] = F(\varepsilon).W_K^I.X[p].W.h_i[p] + z_i[p], i = 1,2,\ldots,M, p = 0,\ldots,P \tag{2}$$

where the normalized frequency offset ε is presented in the matrix $\boldsymbol{F}(\varepsilon)$ given by

$$F(\varepsilon) = \text{diag}\left[1, e^{j(2\pi\varepsilon/K)}, \ldots, e^{j(2\pi\varepsilon(K-1)/K)}\right]_{K\times K} \tag{3}$$

and

$$y_i[p] = \left[y_i[p,0], \ldots, y_i[p,K-1]\right]^T_{K\times 1} \tag{4}$$

$$\left[W_K^I\right]_{r,s} = \frac{1}{\sqrt{K}} e^{j(2\pi rs/K)} (r,s = 0,\ldots,K-1) \tag{5}$$

$$X[p] = [X_1[p], X_2[p], \ldots, X_N[p]]_{K\times N.K} \tag{6}$$

$$X_j[p] = diag\left[x_j[p,0], \ldots, x_j[p,K-1]\right]_{K\times K} \tag{7}$$

$$W = \text{diag}\,[w, \ldots, w]_{NK\times NL} \tag{8}$$

$$[w]_{r,s} = e^{-j(2\pi/K)rs}(r = 0,\ldots,K-1; s = 0,\ldots,L-1) \tag{9}$$

$$h_i[p] = \left[h_{i,1}^T[p],\ldots,h_{i,N}^T[p]\right]_{NL\times 1}^T \tag{10}$$

where $\boldsymbol{h}_i[p]$ is the $NL \times 1$ vector containing the complex channel frequency responses between the ith receive antenna and all transmit antennas at the pth OFDM slot, $x_j[p,k]$ is the symbol transmitted from the jth transmit antenna at the kth subcarrier and at the pth time slot, and$y_i[p,k]$ is the received symbol from the ith receive antenna at the kth subcarrier and at the pth time slot. The additive white Gaussian noise (AWGN) vector $z_i[p]$ has the covariance matrix of $\Sigma_{z_i} = E\left(z_i.z_i^H\right) = \sigma_{z_i}^2.I_K$, where $\boldsymbol{I}_K$ denotes a $K \times K$ identity matrix.

The normalized CFO ε is introduced through the matrix $\boldsymbol{F}(\varepsilon)$ in (2). Since $F(\varepsilon) \neq I_K$ for $\varepsilon \neq 0$, orthogonality among all K subcarriers is destroyed. Due to this loss of subcarrier orthogonality, the interferences from other subcarriers will impact the transmitted symbol at the desired subcarrier. To regain the orthogonality between the MIMO-OFDM subcarriers, we have to correct $\boldsymbol{F}(\varepsilon)$ in (2) before applying the FFT operation. The goal is thus estimating $\boldsymbol{F}(\varepsilon)$ and multiplying the received time domain sequence with $\boldsymbol{F}^{-1}(\varepsilon)$.

3 Frequency Offset and Channel Estimation

To reduce the disturbance effects of CFO, accurate synchronization is important, preferably before reception of the data. Therefore the data packet is preceded by a section of predefined data, which is called the preamble. Our idea is to simultaneously make use of preamble for both CFO and CIR estimations. Therefore, in the sequel to this study, we concentrate on a preamble block (corresponding to $p = 0$) and omit the temporal index p for notational simplicity.

The transmission model in (2) contains two unknown parameters: the CFO ε and the channel parameters $\boldsymbol{h}_i$ (i=1,. . .,M). The ML estimates of ε and $\boldsymbol{h}_i$ are given by minimizing the following quadratic cost function:

$$\min_{\varepsilon,h_i}\left\{\sum_{i=1}^{\mathrm{M}} |y_i - F(\varepsilon)Vh_i|^2\right\} \tag{11}$$

Where

$$V = W_K^I X_{Preamble} W \tag{12}$$

Problem of (11) lies in the fact that we have to estimate two parameters with only one cost function.

To obtain the ML solutions for both frequency offset and channel coefficients, a prohibitive computational complexity is required [21]. Therefore, in [15], the ML estimate for only frequency offset was obtained. In this section, we propose a reduced-complexity scheme for ML estimate of both CFO and CIR. As we shall see, the solution consists of two separate steps: a CFO estimator and a channel estimator. In first step, the CFO ϵ is estimated. In second step, $\hat{\varepsilon}$ is exploited to estimate the MIMO channel coefficients.

3.1 Conventional: ML Estimation of CFO

Sun et al. in [15] devised a new scheme to estimate the CFO in MIMO-OFDM systems. Their proposed scheme performs the CFO estimation as follows:

$$\hat{\varepsilon} = \arg\max_{\varepsilon} \Lambda(\varepsilon) \tag{13}$$

where

$$\Lambda(\varepsilon) = -2\mathrm{Re}\left(\sum_{i=1}^{M} \underbrace{\sum_{n'=1}^{K-1} r_i(n') \mathrm{e}^{-j\left(\frac{2\pi\varepsilon}{K} n'\right)}}_{R_i(\varepsilon)} \right) + \mathrm{const.} \tag{14}$$

in (14)

$$r_i(n') = \sum\nolimits_{m=0}^{K-1-n'} \left[B^H B\right]_{m,m+n'} y_i^*(\mathrm{m}) y_i(m+n') \tag{15}$$

where $B = \left(I_K - (2/K).V.V^H\right)$. Some remarks are of interest as follows:

- Because of the very form of (14), the $R_i^{(k)}(\varepsilon)$ values can be efficiently computed through fast Fourier transform (FFT) techniques. Better frequency resolution can be achieved by zero-padding to effectively increase the length of the preamble sequence.
- As the estimator in (13) requires point search, its complexity depends on the number of points searched over the uncertainly interval. It is possible to reduce the complexity of $\hat{\varepsilon}$ computation in (13) if the search interval restricted to a small interval. This is possible if a coarse estimation module is invoked to bring the CFO to a range manageable by, e.g., [27]. This assumption, together with the FFT-based implementation, will significantly accelerate the search algorithm for computing CFO in (13).
- The computational complexity of estimator can be assessed as follows. Assume that the entries of the matrix $B^H B$have been precomputed. Then, the computation of $\left\{r_i(n'), i = 1, \ldots, M, n' = 1, \ldots, K-1\right\}$ requires a total of $(MK(K-1))$ complex products and $(0.5\,M(K-1)(K-2))$ complex additions. Also, the FFT needs $M(K-1)$ complex products and $M(K-1)-1$ complex additions. The overall operations are summarized in the first row of Table 1, where J denotes the number of points searched over the uncertainly interval.

Table 1. Computational load

Algorithm	Real products	Real additions
Sun	$4MK(K+J-1)-4JM$	$J(4MK-4M-2)+M(3K^2-5K+2)$
Proposed	$IMK(4K+11)$	$4IMK(K+1)-2I$

3.2 Proposal: ML Estimation of CFO and CIR

Step-1: CFO estimator: The ML estimation of ε leads to the following mathematical development:

$$\begin{aligned}\hat{\varepsilon} &= \arg\max_{\varepsilon} \sum_{i=1}^{M} \log p(y_i|\varepsilon) \\ &= \arg\max_{\varepsilon} \sum_{i=1}^{M} \log \int p(y_i|\varepsilon,h_i)p(h_i)dh_i\end{aligned} \tag{16}$$

It is seen in (16) that the direct computation of the optimal ML detection involves multiple-dimensional integral over the unknown random vector $\boldsymbol{h}_i$ and, hence, is of prohibitive complexity. However, it is known that the EM algorithm can provide the ML solutions in an iterative manner for ML estimation problems [25, 26]. Therefore, we resort to the EM algorithm and propose a novel EM-based CFO estimator

The EM algorithm was first introduced by Feder and Weinstein in [26] for parameter estimation of superimposed signals. The idea has been applied in [28] by Xie and Georghiades to channel estimation of OFDM frequency selective fading channels. The basic idea of the EM algorithm is to solve problem (16) iteratively according to the following two steps:

Expectation (E)-step: Compute

$$\mathrm{Q}(\varepsilon|\varepsilon^{(\mathrm{k})}) = \mathrm{E}\left\{\left[\sum_{i=1}^{\mathrm{M}} \log p\left(y_i|\varepsilon,h_i\right)\right] |y_i,\varepsilon^{(\mathrm{k})}\right\} \tag{17}$$

Maximization (M)-step: Solve

$$\varepsilon^{(\mathrm{k}+1)} = \arg\max_{\varepsilon} \mathrm{Q}(\varepsilon|\varepsilon^{(\mathrm{k})}) \tag{18}$$

where $\varepsilon^{(\mathrm{k})}$ denotes the estimated CFO value at the kth EM iteration. It is known that the likelihood function $\sum_{i=1}^{M} \log p\left(y_i\,|\varepsilon^{(\mathrm{k})}\right)$ is non-decreasing as a function of k and under regularity conditions the EM algorithm converges to a local stationary point [29].

In the E-step, the expectation is taken with respect to the hidden channel response $\boldsymbol{h}_{\mathrm{i}}$ conditioned on $\boldsymbol{y}_i$ and $\varepsilon^{(\mathrm{k})}$. It is easily seen that, conditioned on $\boldsymbol{y}_i$ and $\varepsilon^{(\mathrm{k})}$, $\boldsymbol{h}_{\mathrm{i}}$ is complex Gaussian distributed as

$$h_i \Big| \left(y_i,\varepsilon^{(k)}\right) \sim N_c\left(\hat{h}_i,\hat{\Sigma}_{h_i}\right), i=1,\ldots,M \tag{19}$$

with

$$\hat{\Sigma}_{h_i} = \left(V^H F^H(\varepsilon^{(k)}).\Sigma_{z_i}^{-1}.F(\varepsilon^{(k)})V + \Sigma_{h_i}^{\dagger}\right)^{-1} \tag{20}$$

$$\hat{h}_i = \hat{\Sigma}_{h_i}.V^H F^H(\varepsilon^{(k)}).\Sigma_{z_i}^{-1}.y_i \tag{21}$$

where $\Sigma_{z_i}$and Σ_{h_i}denote, respectively, the covariance matrix of the ambient white Gaussian noise z_i and channel responses $\boldsymbol{h}_i$. According to the assumptions in Section 2, both of them are diagonal matrices as

$$\boldsymbol{\Sigma}_{z_i} = E\left(z_i.z_i^H\right) = \sigma_{z_i}^2.I_K \tag{22}$$

$$\boldsymbol{\Sigma}_{h_i} = E\left(h_i.h_i^H\right) = \text{diag}\left[\sigma_{i,j,0}^2,\ldots,\sigma_{i,j,L-1}^2,\ldots,\sigma_{i,N,0}^2,\ldots,\sigma_{i,N,L-1}^2\right] \tag{23}$$

where $\sigma_{i,j,l}^2$ is the average power of the lth tap between the jth transmit and ith receive antennas; $\sigma_{i,j,l}^2 = 0$ if the channel response at this tap is zero. Assuming Σ_{h_i} is known, $\Sigma_{h_i}^{\dagger} = diag[\gamma_{i,1,0},\ldots,\gamma_{i,1,L-1},\ldots,\gamma_{i,N,0},\ldots,\gamma_{i,N,L-1}]$ is defined as the pseudo inverse of Σ_{h_i}as

$$\gamma_{i,j,l} = \begin{cases} \frac{1}{\sigma_{i,j,l}^2} & \sigma_{i,j,l}^2 \neq 0 \\ 0 & \sigma_{i,j,l}^2 = 0 \end{cases} \tag{24}$$

As depicted in Fig. 3, the diagonal elements of $V^H.V$ are equal to *K/N*. *K/N* is much larger than $\sigma_{z_i}^2/\sigma_{i,j,l}^2$, which is inversely proportional to the SNR of the fading channel. Therefore, we can simplify (20) and (21) to

$$\hat{\Sigma}_{h_i} \approx \sigma_{z_i}^2\left(V^H V\right)^{-1} \tag{25}$$

$$\hat{h}_i \approx \left(V^H V\right)^{-1} V^H F^H(\varepsilon^{(k)})y_i \tag{26}$$

Hence, (17) reduces to

$$\begin{aligned} \mathrm{Q}(\varepsilon|\varepsilon^{(\mathrm{k})}) &= -\sum_{i=1}^{M}\left\{E_{h_i|}\left(y_i,\varepsilon^{(\mathrm{k})}\right)\left\{\|y_i - F(\varepsilon)V.h_i\|^2\right\}\right\} + \text{Const.} \\ &= -\sum_{i=1}^{M}\left\{\left\|y_i - F(\varepsilon)V.\hat{h}_i\right\|^2 + trace\left(V\hat{\Sigma}_{h_i}V^H\right)\right\} + \text{Const.} \end{aligned} \tag{27}$$

the second term is independent of ε, and hence has no contribution to the detector. Developing (27) and dropping terms irrelevant of ε, we obtain

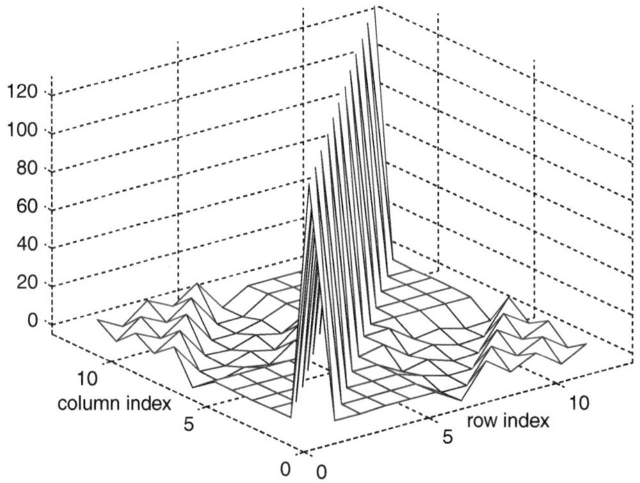

Fig. 3. Evaluation of matrix products $V^H V$ ($K = 256, N = 2$)

$$\begin{aligned} Q(\varepsilon|\varepsilon^{(k)}) &= 2\sum_{i=1}^{M} \operatorname{Re}\left(y_i^H F(\varepsilon)U_i\right) + \text{const.} \\ &= 2\sum_{i=1}^{M}\sum_{n=0}^{K-1} \operatorname{Re}\left(y_i^*(n)U_i(n)exp\left(j\frac{2\pi\varepsilon n}{K}\right)\right) + \text{const.} \end{aligned} \quad (28)$$

where $U_i = V\left(V^H V\right)^{-1} V^H F^H(\varepsilon^{(k)})y_i$. However, it is too complicated to obtain an exact solution to maximize the likelihood function in (28). Therefore, we assume ε is sufficiently small to approximate $\exp(j2\pi\varepsilon n/K)$ by Taylor series expansion to the second-order term

$$\exp(j2\pi\varepsilon n/K) \approx 1 + j(2\pi\varepsilon n/K) - 0.5(j2\pi\varepsilon n/K)^2 \quad (29)$$

then

$$\begin{aligned} Q(\varepsilon|\varepsilon^{(k)}) \approx\, & 2\sum_{i=1}^{M}\sum_{n=0}^{K-1} \operatorname{Re}\left\{y_i^*(n)U_i(n)\right\} - \varepsilon\left(\frac{4\pi}{K}\sum_{i=1}^{M}\sum_{n=0}^{K-1} n\operatorname{Im}\left\{y_i^*(n)U_i(n)\right\}\right) \\ & - \varepsilon^2\left(\frac{4\pi^2}{K^2}\sum_{i=1}^{M}\sum_{n=0}^{K-1} n^2\operatorname{Re}\left\{y_i^*(n)U_i(n)\right\}\right) \end{aligned} \quad (30)$$

Next, based on (30), the M-step proceeds as follows:

$$\varepsilon^{(k+1)} = -\frac{K}{2\pi}\frac{\sum_{i=1}^{M}\sum_{n=0}^{K-1} n\mathrm{Im}\left\{y_i^*(n)U_i(n)\right\}}{\sum_{i=1}^{M}\sum_{n=0}^{K-1} n^2\mathrm{Re}\left\{y_i^*(n)U_i(n)\right\}} \tag{31}$$

The computational complexity of the proposed CFO estimator can be addressed as follows. Assuming $V(V^HV)^{-1}V^H$is precomputed, $MK(K+2)$ complex products and $MK(K\text{-}1)$ complex additions are required to evaluate $U_i$and $\left\{y_i^*(n)U_i(n), 0 \leq \text{n} < K, i = 1,...,M.\right\}$ Moreover, in (31), we need $3MK$ real products and $2(MK-1)$ real additions. Note that a complex product requires four real products and two real additions, whereas a complex addition amounts to two real additions. Then, the overall number of real products and additions can be given as $MK(4K+11)$ and $4MK(K+1)-2$, respectively. The second row in Table 1 shows the computational load involved in the search of the propose CFO estimator, where I denotes the number of needed EM iterations.

Step-2: Channel estimator: When $\hat{\varepsilon}$ has been obtained, it is straightforward, using (11), to estimate $\boldsymbol{h}_i$. The LS solution of (11) is also the ML channel estimate. The ML solution can be determined by

$$\tilde{h}_i = (V^H.V)^{-1}.V^H F^H(\hat{\varepsilon}).y_i \tag{32}$$

4 Performance Analysis

In this section to check the optimality of the proposed scheme we derived the CRBs for both CFO and CIR estimators.

Remember that for each received antenna we have $y_i = F(\varepsilon)V.h_i + z_i (i = 1,...,M)$.Stacking all the different vectors $\boldsymbol{y}_i$ in column to form the vector $\boldsymbol{y}$ and doing the same thing with vectors $\boldsymbol{h}_i$ to obtain $\boldsymbol{h}$, we add together the contributions of all received antennas

$$\begin{bmatrix} y_1 \\ \vdots \\ y_M \end{bmatrix} = \begin{bmatrix} F(\varepsilon).V & & 0 \\ & \ddots & \\ 0 & & F(\varepsilon).V \end{bmatrix} \times \begin{bmatrix} h_1 \\ \vdots \\ h_M \end{bmatrix} + \begin{bmatrix} z_1 \\ \vdots \\ z_M \end{bmatrix} \tag{33}$$

or in the matrix form

$$y = (I_M \otimes F(\varepsilon)V).h + z \tag{34}$$

where $\otimes$ denotes the Kronecker product, and $\boldsymbol{z}$ is a noise vector which is classically assumed to be zero-mean circularly symmetric complex-valued Gaussian with covariance matrix $\Sigma_z = \sigma_z^2.I_{MK}$.

Let $\boldsymbol{\eta} = [\varepsilon h_R h_I]^T$ denote the parameter vector of interest where $\boldsymbol{h}_R$ and $\boldsymbol{h}_I$ stand for the real and imaginary parts of $\boldsymbol{h}$, respectively. Under all the made assumptions, the received signal $\boldsymbol{y}$ is a complex-valued circularly symmetric Gaussian vector with mean $\mu = (I_M \otimes F(\varepsilon)V) .h$ and covariance matrix $C_z = \sigma_z^2.I_{M.K}$. For this type of problem, the Fisher Information Matrix (FIM) for estimation of $[\eta^T \sigma_z^2]$ is block diagonal, i.e., the estimation of σ_z^2 is decoupled from that of η. Therefore, we only consider the FIM for η, which we denote by $\boldsymbol{F}$. The latter is given by [30]

$$F = \frac{2}{\sigma_z^2} Re(\frac{\partial \boldsymbol{\mu}^H}{\partial \boldsymbol{\eta}} \frac{\partial \boldsymbol{\mu}}{\partial \boldsymbol{\eta}^T}) \tag{35}$$

The CRB is obtained as the inverse of the FIM $\boldsymbol{F}$. Using the same principle as those given in [30], we obtain the final expression for CRB as

$$CRB(\varepsilon) = \frac{K^2\sigma_z^2}{8\pi^2} \left[h^H \left(I_M \otimes V^H D \boldsymbol{\Pi} DV\right) h\right]^{-1} \tag{36}$$

with the following definitions:

$$D = diag(0,1,...,K-1) \tag{37}$$

$$\boldsymbol{\Pi} = I_K - V(V^H V)^{-1} V^H \tag{38}$$

Using the same approach of [30], we obtain for any unbiased estimate $\hat{h}$ of $\boldsymbol{h}$

$$E[(\hat{h} - h)(\hat{h} - h)^H] \geq 0.5\sigma_z^2(2\lambda + \gamma^{-1}\beta\beta^H) = CRB(h) \tag{39}$$

where

$$\lambda = I_M \otimes \left(V^H V\right)^{-1} \tag{40}$$

$$\gamma = h^H I_M \otimes \left[V^H D \boldsymbol{\Pi} DV\right] h$$

$$\beta = I_M \otimes \left[\left(V^H V\right)^{-1} V^H DV\right] h$$

As indicated in (36) and (39), CRB values depend on the specific channel realization $\boldsymbol{h}$.

5 Simulation Results

In this section, we provide computer simulation results to illustrate the performance of the proposed scheme. The characteristics of the fading channels are described in Section 2, specifically; the system performance is simulated in typical urban (TU) channel with six equal-power taps. In the following simulations, the available bandwidth is 1 MHz and is divided into 128 subcarriers. These correspond to a subcarrier symbol rate of 7.8 kHz and OFDM word duration of 128 μs. In each OFDM word, a cyclic prefix interval of 32 μs is added to combat the effect of ISI; hence, the duration of one OFDM word is 160 μs. For all simulations, two transmitter antennas and two receiver antennas are used. The information symbols are drawn from a quaternary phase-shift keying (QPSK) constellation. The simulated system transmits data in a burst manner. Each data burst includes 10 OFDM blocks. A preamble is applied at the beginning of each data burst for synchronization purposes. MIMO channel estimates are also drawn from the preamble. The performance of CFO estimator is evaluated with the mean-square error (MSE) of the estimated frequency offset. In the case of channel estimation parameters, $E\left(\left\|h-\hat{h}\right\|^2\right)$ is plotted.

Example 1 (performance of CFO estimator): First, we test the effect of the number of EM iterations on CFO estimation. The CFO is randomly selected in the range $[-0.1, 0.1]$. To start the iteration of proposed algorithm, we set the initial estimate $\varepsilon^{(0)} = 0$. In Fig. 4, the MSE performance of the CFO estimator versus the number of EM iterations is depicted. As it can be seen the MSE performance of the proposed estimator converges after three iterations at the SNR of 5 dB, while it is improved continually until the number of iterations increases to 12 for the SNR values greater than 20 dB.

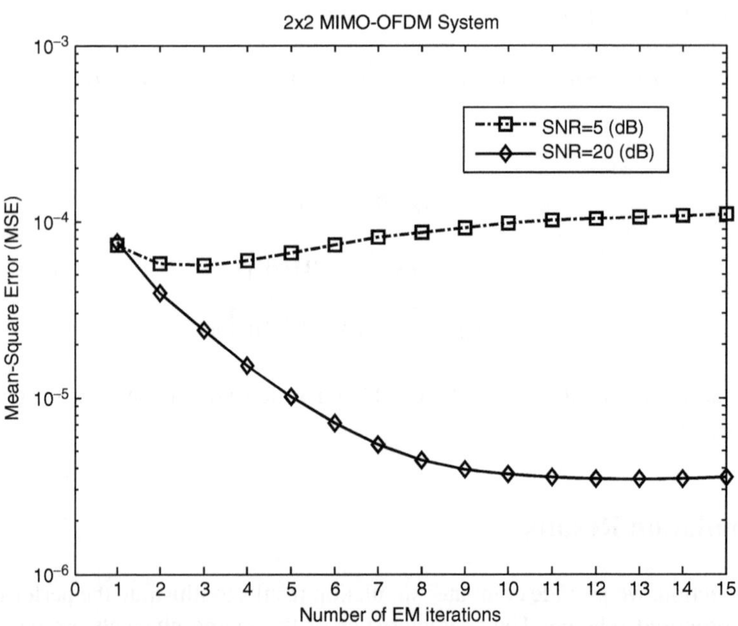

Fig. 4 MSE performance of the CFO estimator versus number of EM iterations for a 2×2 MIMO-OFDM communication system

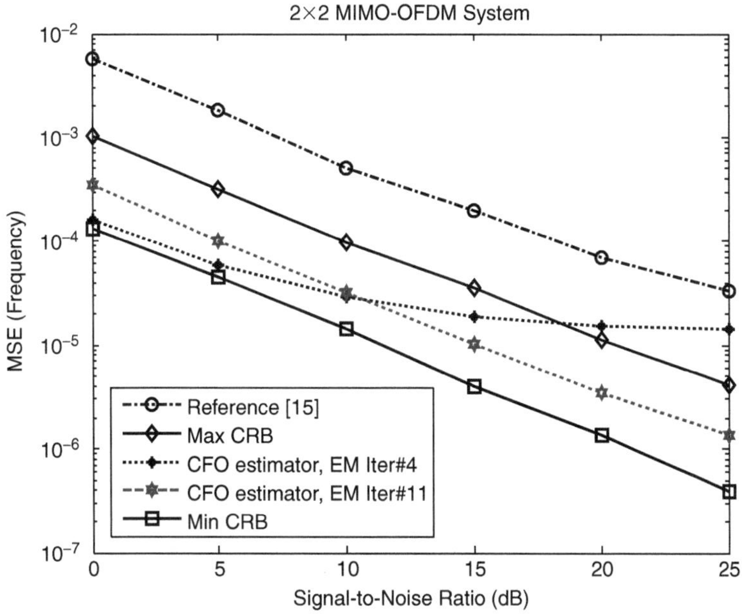

Fig. 5 MSE of CFO estimation versus SNR for a 2×2 MIMO-OFDM communication system

In Fig. 5, we depict the MSE performance versus SNR. The curves denoted by "EM Iter#4" and "EM Iter#11" show the CFO estimator performance after the 4th and 11th EM iteration, respectively. The CRB derived in (36) is also shown as a bench mark. As indicated in (36), CRB values are varying according to the channel state [18], and for a given SNR, the solid lines labeled "Min/Max CRB" indicate the minimum and maximum CRB obtained in 10^5 simulation runs. However, in high SNR above 20 dB, the proposed scheme needs 11th iterations to maintain the improved performance between CRBs in all ranges of SNR. From Fig. 5, it is also deduced that proposed frequency estimator outperforms the one in [15] in all ranges of SNR. Note that for Sun et al. method, we search over 2,000 points equispaced within the range $[-0.1, 0.1]$.

Figure 6 compares the total required operations (real additions plus real multiplications) of Sun et al. CFO estimator [15] with our EM-based CFO estimator, as a function of the number of subcarriers. The curves are computed from the results in Table 1 with $J = 2000$, $I = 10$, $M = 2$. It is clearly seen that relative to Sun et al. estimator [15], our EM-based CFO estimator has lower computational complexity.

Example 2 (performance of channel estimator): In this example, we test the performance of MIMO channel estimation with *(N,M)* = *(2,2)* and CFO being also randomly selected in the range $[-0.1, 0.1]$. In Fig. 7, the simulated MSE performances of channel estimator in (32) are presented and compared with the *CRB* in (39) and "ideal case," where the CFO is perfectly known. It is seen that the channel estimator of the proposed algorithm shows almost ideal performance in all ranges of

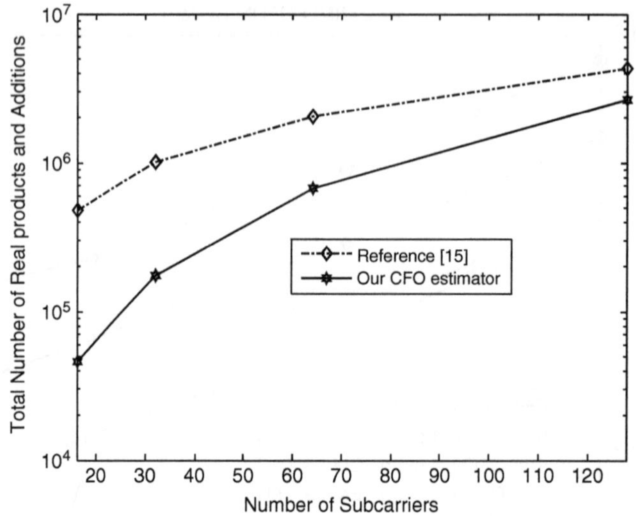

Fig. 6 Computational complexity of proposed CFO estimator

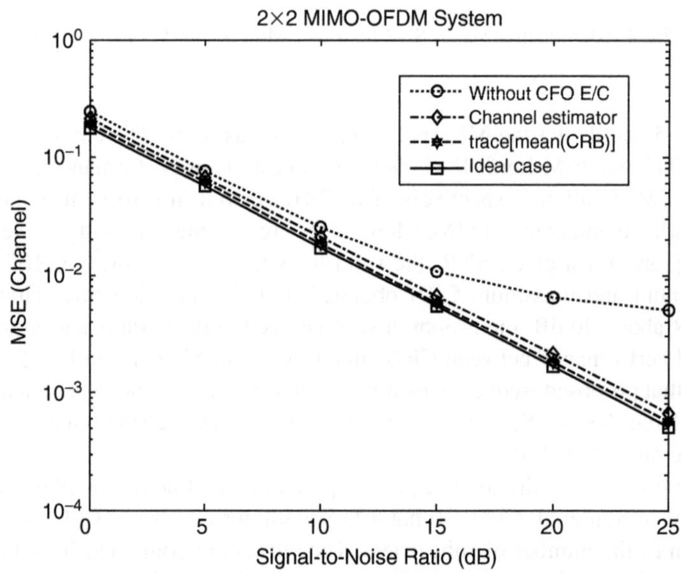

Fig. 7 Channel estimation performance of a 2×2 MIMO-OFDM communication system

SNR. To study the disturbance effects of CFO on channel estimation performance, we also plot the MSE of channel estimator without estimating/compensating CFO, in curve denoted by "Without CFO E/C." It is not surprising to see the significant performance loss due to CFO.

6 Conclusions and Future Work

The problem of estimating the frequency offset and channel coefficients in multi antenna OFDM transmission was investigated in this chapter. The frequency estimates were obtained using an ML approach, and the EM algorithm was employed to reduce the computational complexity of ML solution. The CFO estimates were then exploited to estimate the MIMO channel responses. The performance of our estimators was benchmarked with CRBs and investigated by computer simulation. Simulation results show that the proposed algorithm achieves almost ideal performance compared with the CRBs for both channel and frequency offset estimations.

The possible directions for the future research include verifying the performance of our frequency offset and channel estimators for different space–time techniques such as the Bell labs layered space–time (BLAST) [4], use of powerful coding schemes such as the low-density parity check (LDPC) and the turbo codes [31], and finally, extending the algorithm to the multi-user scenarios.

References

1. A. Wittneben, "A new bandwidth efficient transmit antenna modulation diversity scheme for linear digital modulation," in *Proc. IEEE Int. Communications Conf.*, Geneva, Switzerland, June 1993, pp. 1630–1634.
2. V. Tarokh, N. Seshadri, and A. R. Calderbank, "Space-time codes for high data rate wireless communication: Performance analysis and code construction," *IEEE Trans. Inform. Theory*, 44, 744–765, Mar. 1998.
3. J. H. Winters, "On the capacity of radio communication systems with diversity in a Rayleigh fading environment," *IEEE J. Select. Areas Commun.* , SAC–5, 871–878, June 1987.
4. G. J. Foschini, "Layered space–time architecture for wireless communication in a fading environment when using multi-element antennas," *Bell Labs Tech. J.*, 22, 41–59, Autumn 1996.
5. G. J. Foschini and M. J. Gans, "On limits of wireless communications in a fading environment when using multiple antennas," *Wireless Personal Commun.*, 6, 311–335, 1998.
6. G. J. Foschini, G. D. Golden, R. A. Valenzuela, and P. W. Wolniansky, "Simplified processing for high spectral efficiency wireless communication employing multi-element arrays," *IEEE J. Select. Areas Commun.*, 17, 1841–1852, Nov. 1999.
7. G. L. Stuber, S. W. Mclaughlin, Y. Li, M. Ingram, and T. G. Pratt, "Broadband MIMO-OFDM Wireless Communications", *IEEE Proc.*, 92, 2, 271–294, Feb. 2004.
8. Y. Li, J. H. Winters, and N. R. Sollenberger, "MIMO-OFDM for wireless communications, signal detection with enhanced channel estimation", *IEEE Trans. Commun.*, 50, 9, 1471–1477, Sep. 2002.
9. H. Bolckei, D. Gesbert, and A. Paulraj, "On the capacity of OFDM based spatial multiplexing systems", *IEEE Trans. Commun.*, 50, 225–234, Feb. 2002.
10. M. Engels, *Wireless OFDM Systems: How to make them work?* Newton, MA: Kluwer Academic Publishers, 2002.
11. B. Lu, X. Wang and K. R. Narayanan, "LDPC-Based space-time coded OFDM systems over correlated fading channels: performance analysis and receiver design," *IEEE Transactions on Communication*, 50, 1, 74–88, Jan 2002.
12. X. Ma, M. K. Oh, G. B. Giannakis, and D. J. Park, "Hopping pilots for estimation of frequency-offset and multiantenna channels in MIMO-OFDM," *IEEE Trans. Commun.*, 53, 1, 162–172, 2005.

13. T. C. W. Schenk and A. van Zelst, "Frequency synchronization for MIMO OFDM wireless LAN systems," *in Proceedings of the IEEE VTC*, Orlando (FL), Oct. 2003, pp. 781–785.
14. P. Priotti, "Frequency synchronization of MIMO OFDM systems with frequency selective weighting," *in Proceedings of the IEEE VTC*, 2004, 2, 1114–1118.
15. Y. Sun, Z. Xiong, and X. Wang, "EM-based iterative receiver design with carrier frequency offset estimation for MIMO OFDM systems," *IEEE Trans. Commun.*, vol. 53, no. 4, pp. 581–586, Apr. 2005.
16. Z. J. Wang, Z. Han, and K. J. Ray Liu, "A MIMO-OFDM channel estimation approach using time of arrivals," *IEEE Trans. Wireless Commun.*, vol. 4, no. 3, pp. 1207–1213, 2005.
17. H. Minn, and N. Al-Dhahir, "Optimal training signals for MIMO-OFDM channel estimation," *IEEE Trans. Wireless Commun.*, vol. 5, no. 5, pp. 1158–1168, 2006.
18. M. Morelli, and U. Mengali, "Carrier-frequency estimation for transmissions over selective channels", *IEEE Trans. Commun.*, vol. 48, no. 9, pp. 1580–1589,2000.
19. X. Ma, H. Kobayashi, and S. C. Schwartz, "Joint frequency offset and channel estimation for OFDM", *Proc. IEEE Global Telecom. Conf.*, 2003, vol. 1, pp. 15–19.
20. T. Cui and C. Tellambura, "Robust joint frequency offset and channel estimation for OFDM systems", *Proc. IEEE VTC Conf.*, Los Angeles, CA, 004, vol. 1, pp. 603–607.
21. M. O. Pun, M. Morelli, and C. C. Jay Kuo, "Maximum-likelihood synchronization and channel estimation for OFDMA uplink transmissions", *IEEE Trans. Commun.*, vol. 54, pp. 726–736, 2006.
22. I. Ziskind and M. Wax, "Maximum likelihood localization of multiple sources by alternating projection", *IEEE Trans. Acoustic, Speech, Signal Process.*, vol. 36, no. 10, pp. 1553–1560, 1988.
23. G. J. McLachlan and T. Krishnan, *The EM algorithm and extensions,* John Wiley and Sons, New York, 2000.
24. T. Moon, "The expectation-maximization algorithm,"*IEEE Signal Processing Magazine*, vol. 13, pp. 47–60, 1996.
25. C. N. Georghiades and J. C. Han, "Sequence estimation in the presence of random parameters via the EM algorithm," *IEEE Trans. Commun.*, vol. 45, no. 3, pp. 300–308, 1997.
26. M. Feder and E. Weinstein, "Parameter estimation of superimposed signals using the EM Algorithm,"*IEEE Trans. ASSP*, vol. 36, no. 4, pp. 477–489,1998.
27. S. A. Fechtel, "OFDM carrier and sampling frequency synchronization and its performance on stationary and mobile channels," *IEEE Trans. Consum. Electron*, vol. 46, pp. 438–441, 2000.
28. Y. Xie and C. N. Georghiades, "Two EM-type channel estimation algorithms for OFDM with transmitter diversity," *IEEE Trans. Commun.*, vol. 51, no.1, pp. 106–115, Jan. 2003.
29. H. V. Poor, *An Introduction to Signal Detection and Estimation*, 2nd ed., Springer-Verlag, New York, 1994.
30. P. Stoica and O. Besson, "Training sequence design for frequency offset and frequency selective channel estimation"*IEEE Trans. Commun.*, vol. 51, pp. 1910–1917, Nov. 2003.
31. C. Berrou and A. Glavieux, "Near optimum error correcting coding and decoding: turbo codes," *IEEE Trans. Commun.*, vol. 44, pp. 1261–1271, Oct. 1996.

Wireless Communication Systems from the Perspective of Implantable Sensor Networks for Neural Signal Monitoring

C. Tarín, L. Traver, P. Martí, and N. Cardona

Abstract Recent advances in modern neurocomputing heading toward promising clinical applications of implantable neuronal sensing devices have shown the utmost necessity of wireless communication systems that allow real-time monitoring of neural signals. The design of a wireless transmission system for this particular application shall meet several requirements involving source compression of the high data rate neural recording, communication with a standard device as bridge between body area and remote server, and high fidelity of the received signal to ensure effective brain activity monitoring. A wireless transmission system over Bluetooth and 3G is analyzed for its application to the real-time transmission of neural signals captured by implanted micro-electrode array sensors. Average compression rate of 75% of the neural signal is achieved through detection using nonlinear energy operator preprocessing and automatic threshold adaptation. The wireless transmission of these signals integrates a Bluetooth transmission from the information source to a conventional mobile device and then over 3G to a remote server, without intermediate storage on the mobile phone. Reconstruction of the coded neural signal provides the input to high-performance spike classification algorithm allowing the tracking of individual neuron spiking patterns.

1 Introduction

In recent years, a number of promising clinical prototypes of implantable and wearable monitoring devices have started to emerge [32]. Although a number of problems as long-term stability and biocompatibility remain, the potential medical value is enormous. Many applications exist in the field of bio-telemetry: blood glucose

C. Tarín (✉)
Institute for Telecommunications and Multimedia Applications, Technical University of Valencia, Valencia, Spain
e-mail: salari@eetd.kntu.ac.ir

S. Powell, J.P. Shim (eds.), *Wireless Technology*, Lecture Notes in Electrical Engineering 44, DOI 10.1007/978-0-387-71787-6_12,

level monitorization, identification in cardiological life-threatening episodes, etc. Our interest focuses in neural signal recording and monitoring [4, 30].

Studies conducted during the last decade have demonstrated the improvements that neural signal decoding will bring to health care, especially for patients suffering from paralysis [6], blindness [5], or deafness [22]. Spike processing techniques and, in particular, spike detection and classification are fundamental in analyzing and interpreting both in vivo and in vitro recordings of neural activity. Basic spike detection algorithms apply threshold-based detection to identify spikes and, although simple thresholding is attractive for real-time implementations because of its computational simplicity, it is thought to be sensitive to noise and requires user input to set effective threshold levels [2].

Telemetry systems for neuronal signals are nowadays under investigation. In fact, the implementation of a wireless transmission method for such systems brings considerable advance especially for in vivo recordings as the subject wearing the measurement devices would then be freely moving around and neural recordings from normal life-style activities would be available. Wireless-implanted neural electrodes that have been characterized and tested in vivo are reported in [1, 7, 15]. Heading one step further, the involved wireless transmission system should be such that a conventional portable device could figure as bridge between the real-time neural recording of the implanted device and a remote server where data analysis is performed. Thus, Bluetooth and third-generation mobile communication (3G) are particularly interesting as possible wireless transmission methods for such systems as these techniques are available on conventional mobile phones and other portable devices [13, 32].

The design of a telemetry application includes the specification of the communication system parameters such as data transmission capacity, synchrony, and delay. In other words, the application defines the set of requirements to be met by the communication system [23]. Meeting the required specifications is eventually accomplished by correct definition of the communication protocol layers which use a physical channel to implement the data communication.

We have developed a wireless transmission system for neural signals over Bluetooth and 3G. The neural signals are recorded by micro-electrode arrays and then, in real-time, transmitted over a Bluetooth link to a mobile phone. This mobile device immediately, without intermediate storage, re-transmits the signals over 3G to a remote server where data processing and analysis is performed.

First, in Section 2, there is a description of the characteristics of the signals captured by the neuronal multi-electrode sensors and the bandwidth requirements imposed by them. This section also includes a compression algorithm based on spike detection. Detection quality is improved through an adaptive threshold method to be used with nonlinear energy operator spike detection and results are presented by means of receiver operating curves.

Section 3 describes and analyzes the developed wireless transmission system over Bluetooth and 3G. This section includes an analysis of the transmission capability depending on the source compression factor. Reconstruction algorithms and signal post-processing are detailed in Section 4, where the focus lies on the classification algorithms that allow the sorting of the recorded spikes and thus, the extraction of relevant neural activity parameters.

Finally, in Section 5 the conclusions drawn from the developed work are shown and the future trends are discussed providing insight about the future of the wireless communication systems from the perspective of implantable sensor networks for neural signal monitoring.

2 Neural Signal Processing

Neural signals recorded either from in vitro cultures or from in vivo subjects present a high data rate information source. Due to the limited bandwidth of wireless transmission system, compression becomes mandatory. Compression through spike detection becomes extremely attractive when aiming at real-time applications and individual neuron spike pattern analysis.

2.1 Neural Signals

Signals from extracellular cortical electrodes contain action potential waveforms with amplitudes ranging from tens to hundreds of microvolts peak to peak; pulse widths are typically 1–1.5 msec. The noise floor, which includes biological noise from far field neurons and electrical noise from the amplifier circuit, is around 20 μVrmsec; signal to noise ratios (SNRs) therefore range from 0 to 12 dB, although ratios as high as 20 dB are occasionally encountered. Published figures for the signal frequency content vary, ranging from 100 to 400 Hz for the low-end range and 3 k to 10 kHz for the high-end range [18].

Published sampling rates also vary, ranging from 15 up to 50 kHz [20]. In general, higher sampling rates produce higher fidelity signals but also produce more data, requiring faster and higher-power systems to process them, which may become a handicap in a wireless system with limited bandwidth.

Analog-to-digital converter (ADC) resolution should be 10–12 bits to provide 60–72 dB of dynamic range.

In general, the required transmission bandwidth ($Bwth$) can be obtained as $Bwth = fs \cdot nbits \cdot Nch$, where fs is the sampling frequency in samples per second, $nbits$ the number of bits per sample, and Nch the number of channels to be transmitted. Figure 1 displays the data rate required for the transmission of a neural signal as a function of the sampling frequency and the selected number of quantization bits per sample. As it can be observed from this figure, for the transmission of one single channel with a medium sampling rate the required bandwidth is of 180 kbps. When it comes to simultaneous transmission of several channels (up-to-date in vivo and in vitro micro-electrode array recording systems provide up to 128 simultaneous channels [19]) including source compressing algorithms in the communication

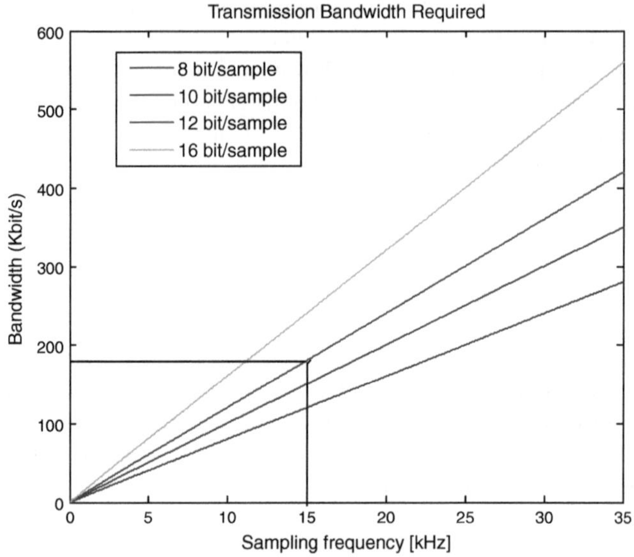

Fig. 1 Transmission bandwidth required depending on the sampling frequency and precision

system becomes a must, especially when wireless transmission with limited bandwidth availability is aimed at.

The spiking activity among the recorded channels might differ substantially, and therefore, as widely discussed in literature [10], spike detection yields the most suitable compression algorithm. Quantitatively assessing detection quality [20, 31] requires knowledge of the ground truth, i.e., decisions taken by the algorithm on the presence of spikes must be compared with the real presence of spikes in the signal. Recordings from micro-electrode arrays do not allow intra-cellular recording which means that the ground truth is not known.

In order to overcome this problem we select two different types of source signals.

First, recordings of in vitro neural activity kindly provided by multichannel systems. A 64-electrode array was used and signals where sampled at 15 kHz. Figure 2 shows the multichannel systems multi-electrode array (MEA). One channel of the recording was selected and spikes were manually detected by several experts. Manual detections were used as the ground truth for the evaluation of detection algorithms.

Second, we use a set of synthetic signals from a statistical model resembling real signals, where the spike positions are known and can be used for the evaluation of spike detection algorithms.

The set of artificial signals contains 10 different signals resulting from adding an artificially generated neuronal noise with a principal neuron spike train. We started with in vivo recordings from rat cerebellum's striatum cells, publicly available at [28]. From these recordings we isolated 50 action potentials and an additional one was selected and repeated periodically with a frequency of 50 Hz to

Fig. 2 64 channels multi-electrode array (MEA) from multichannel systems (www.multichannelsystems.com) widely used in neuronal signal recordings

construct the principal neuron spike train of 10 sec duration. To generate a realistic underlying noise, we assumed that each neuron fires according to a homogeneous Poisson process. The Poisson model is valid if one assumes that each neuronal spike constitutes an independent random variable, which is not totally true but it yields to an approximation that suffices for the generation of additive noise, where the importance is not on the exact spiking times but on the fact that the resulting noise resembles the real neuronal noise present in micro-electrode recordings. The number of noise neurons taken for noise generation is an approximation based on the assumptions that only neurons within 140 μm of the electrode are detectable and that the density of the motor cortex neurons is 30,000 neurons/mm [2]. A scheme of the procedure is shown in Fig. 3. First, the firing rate for each neuron is obtained randomly in the range [50, 90] Hz, then, a firing pattern for each neuron is obtained using the Poisson process model, and finally, the resulting noise is the sum of the individual firing patterns. The principal neuron spike train is added to the adequately attenuated noise to obtain 10 different signals with SNRs in the range [1, 4.6] dB.

2.2 Neural Signal Compression

As described above, neural signals contain trains of action potentials or spikes that form particular spiking patterns. During the intervals of the signal without spikes, the content of the signal is exclusively noise. It is, therefore, possible to compress neural signals by coding the impulse trains leaving the noise-only parts away. For doing this, it is necessary first to detect the occurrence of the spikes, and then code the time, the channel (in the case of a multichannel recording system), and the spike waveform. In this way, it is feasible to compress and multiplex an arbitrary number of channels into one single stream of data. For this work, we have considered the multichannel systems MEA case where the recording system has

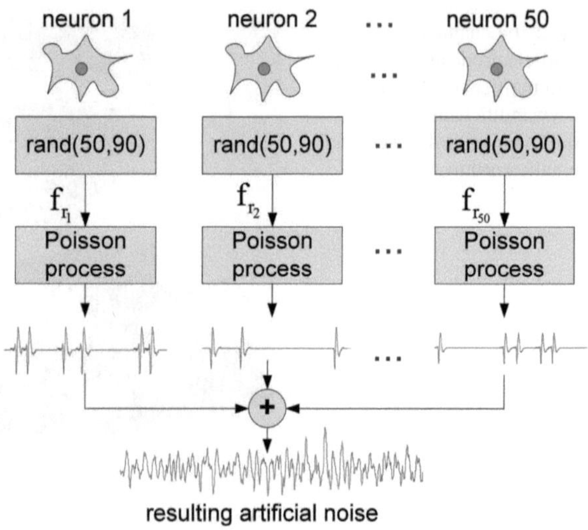

Fig. 3 Artificial noise-generation process

64 recording electrodes, with the sampling frequency being 15 kHz and the sampling precision 12 bits. The system is then producing a $12 \times 15 \times 64 = 11{,}520$ kbps data stream.

We have implemented a compression algorithm that works in a frame-based manner. The algorithms takes input data samples in frames containing 750 samples, i.e., 50 ms. In each frame-based step the algorithm performs spike detection for each of the channels and, when a spike is detected, the time and channel of the spike are coded at the output. Figure 4 shows the coding structure. Each coded spike results in 78 output samples to be transmitted: 1 for the channel, 2 samples for the coding the time-stamp, and 75 samples corresponding to the spike waveform.

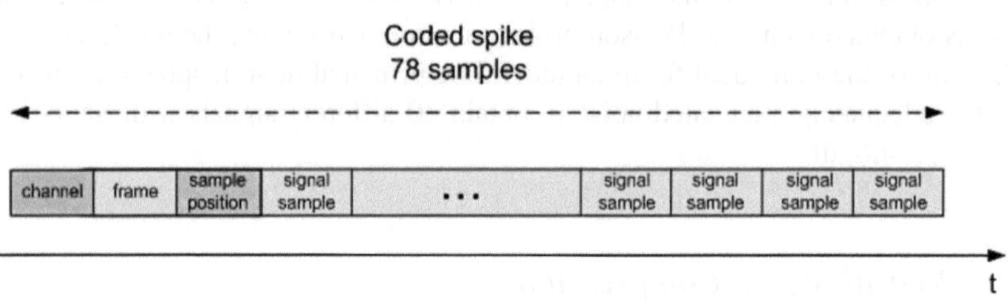

Fig. 4 Coded spike

Neural signals coded in this way can be later decoded and spike trains can be reconstructed by placing each spike waveform in the corresponding channel and time. Detection is done by a nonlinear energy operator (NEO)-based detector.

Basically the algorithm performs a preprocessing stage before detection which consists of the application of the following nonlinear operation on the input signal s[n]:

$$\Psi\,(s[n]) = s^2[n] - s[n-1] \cdot s[n+1] \qquad (1)$$

After preprocessing, spikes are detected comparing the preprocessed signal with an adaptive threshold. Real-time adaptation is done by obtaining a noise-envelope estimate via a frame-based noise-envelope tracking method and, then, setting the threshold $th[\mathrm{k}]$ to a certain level which is proportional to the estimated noise-envelope n[k] (proportionality factor C), where k is the frame number:

$$th[\mathrm{k}] = C \cdot \mathrm{n}[\mathrm{k}] \tag{2}$$

Such preprocessing eases the detection process because it amplifies the signal energy concentrations.

After preprocessing, spikes are detected by comparing the resulting signal amplitude with an adaptive threshold. Adaptation is done by performing automatic noise-level tracking and setting the threshold to a certain level which is relative to the estimated noise level.

As it is exposed in [2], signal-to-noise ratios (SNRs) varies with electrode geometry, size, and position with respect to the target neuron. That is why it is necessary to individually set the threshold to the appropriate value and the adaptive threshold setting becomes useful.

The process for noise-level estimation is as follows. For each processing frame:

1. Calculation of the maximum absolute value of the signal amplitude $|\mathrm{s}_{\max}[n]|$.
2. Comparison of the maximum with the noise-level estimation in the previous frame $|\mathrm{n}[\mathrm{n}-1]|$.
3. If the maximum is bigger than C times the noise level of the previous frame it is assumed that there is a spike present in the frame and therefore, the noise-level estimate is not updated.
4. Otherwise,

   a. if the maximum is bigger than the noise-level estimate, then the noise level is updated through augmentation:

$$|\mathrm{s}_{\max}[\mathrm{n}]| > |\mathrm{n}[\mathrm{n}-1]| \Rightarrow |\mathrm{n}[\mathrm{n}]| = \alpha_{\mathrm{up}} \cdot |\mathrm{s}_{\max}[n]| + \left(1-\alpha_{\mathrm{up}}\right) \cdot |\mathrm{n}[\mathrm{n}-1]| \tag{3}$$

   b. if the maximum is smaller than the noise-level estimate, then the noise level is updated through reduction:

$$|\mathrm{s}_{\max}[\mathrm{n}]| < |\mathrm{n}[\mathrm{n}-1]| \Rightarrow |\mathrm{n}[\mathrm{n}]| = \alpha_{\mathrm{dw}} \cdot |\mathrm{s}_{\max}[n]| + (1-\alpha_{\mathrm{dw}}) \cdot |\mathrm{n}[\mathrm{n}-1]| \tag{4}$$

Time constant values have been experimentally adjusted yielding adequate noise-level tracking. An example of automatic noise-level tracking is shown in Fig. 5.

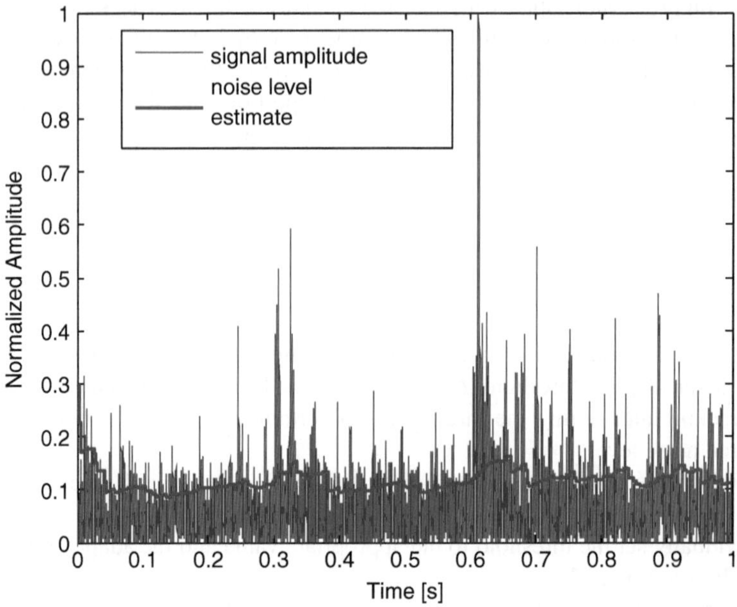

Fig. 5 Automatic neural signal noise-level tracking example

2.3 Compression Results

NEO-based detector with automatic noise-envelope tracking algorithm is used to detect spikes in the set of real and artificial signals. Figure 6 shows spike detections given by NEO detector with automatic noise-envelope tracking for different real recording channels that offer different signal-to-noise ratios. One can observe that threshold adaptation to the appropriate level above the underlying noise occurs in about 0.5 sec and that it is not affected by the spiking activity. Comparison among the four tested channels (Fig. 6a–d) shows the algorithm ability to adapt to different SNR conditions.

To evaluate the detection performance, receiver operating curves (ROCs) have been plotted from the spike detection results. Figure 7 shows ROC families obtained for NEO when applied to the artificial signals set. It also includes resulting probabilities of detection and false alarm for the different SNRs obtained using the adaptive threshold method. Here one can see that the adaptation mechanism sets the detection working point according to the input SNR. Arrows in Fig. 7 indicate the moving direction of the working point with changing SNR if a fix threshold would be used with the consequent performance degradation.

Similarly, Fig. 8 plots ROCs corresponding to the real signal and shows that NEO curve is close to the ideal detection curve, which is the step function. Area-under-curve figure for NEO is 0.9473.

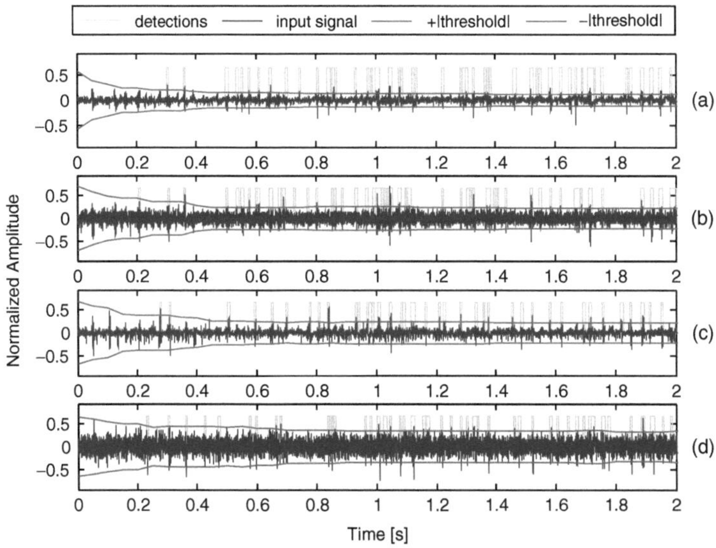

Fig. 6 Automatic neural signal noise-envelope tracking for different real recording channels with different signal to noise ratios. Input signal is plotted in *blue*, detections are marked in *yellow*, and the threshold level is depicted using *magenta*

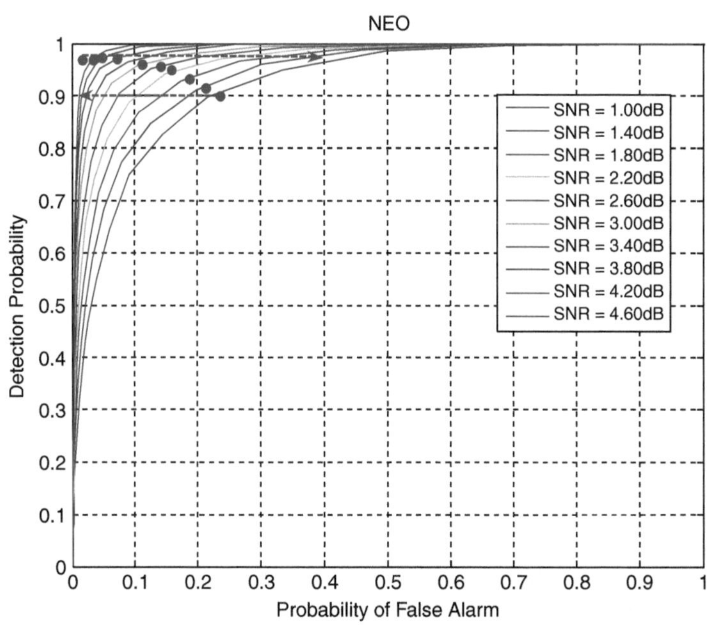

Fig. 7 Family of ROC curves obtained for the set of 10 artificial signals with SNRs ranging from 1 to 4.6 dB. *Thick dots* on of the curves correspond to the detection and false alarm probabilities obtained using adaptive threshold. *Arrows* indicate working point moving direction if a fix threshold is used for the rest of SNR conditions

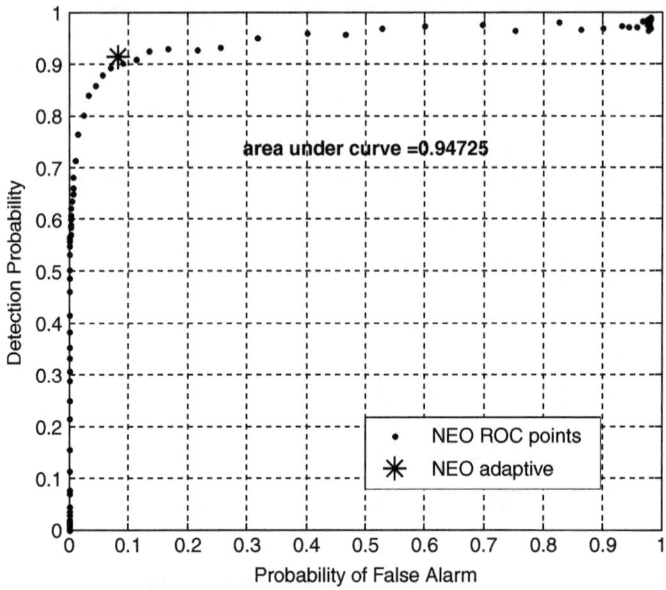

Fig. 8 ROC curve obtained using real data for NEO. Asterisk on *top* of the curve corresponds to the detection and false alarm probabilities obtained using adaptive threshold

3 Wireless Transmission

Figure 9 represents the overall transmission scheme. The information source is a personal computer (**PC**), where neural data recorded by the multichannel systems MEA are stored. This PC establishes via a Bluetooth-Dongle a wireless communication link with a mobile terminal. The stored data are preprocessed as described in Section 2 and transmitted over the Bluetooth link from the information source to the mobile device that receives them and, without intermediate storing, re-transmits them via a 3G link to a remote server PC, where the data are definitely stored, reconstructed, and post-processed.

3.1 Bluetooth Wireless Transmission

Bluetooth is a flexible and capable technology for providing short-range radio communications between devices in an ad hoc manner using the 2.4 GHz band. It is well suited as a low-power radio transceiver (transmitter and receiver) operating at up to 1 Mbps [3]. Two types of channels are used in Bluetooth systems: SCO and ACL. SCO are synchronous connection-oriented links with fixed 64 kbps data rate used exclusively for voice traffic; while ACL are asynchronous connection-less links. As shown above in Section 2, streaming of multichannel or even single channel neural signals demands such a bandwidth which cannot be offered by SCO links.

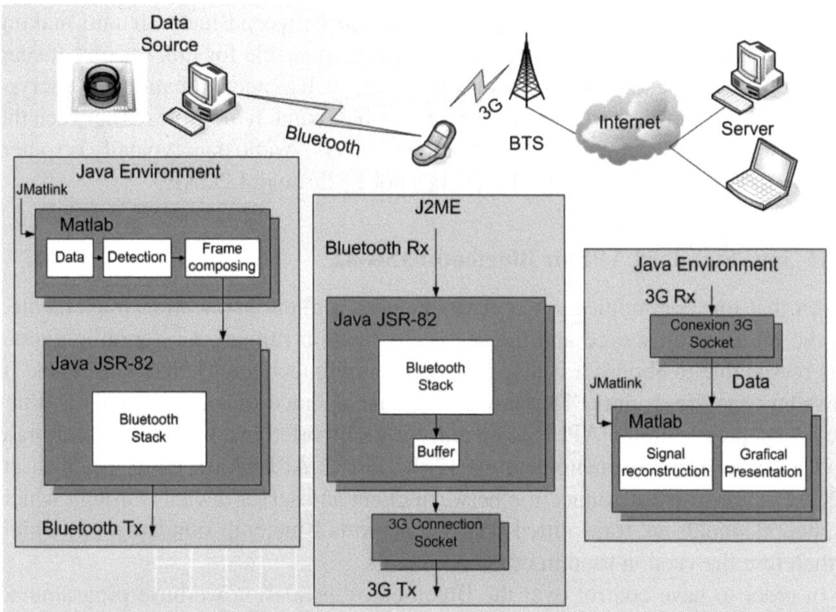

Fig. 9 Wireless Bluetooth-3G transmission

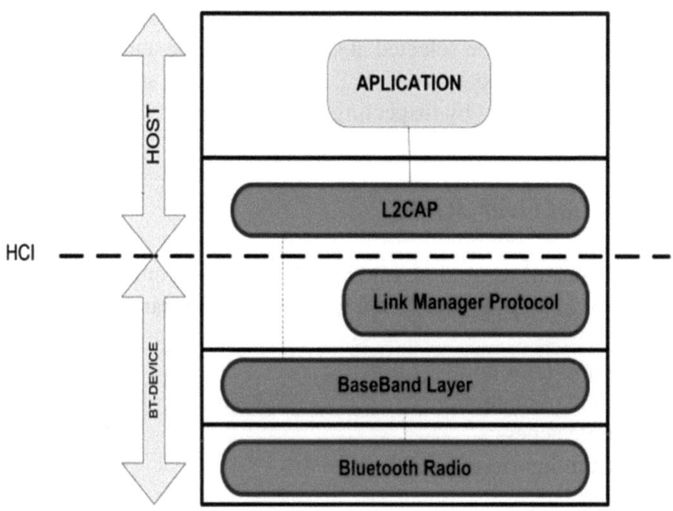

Fig. 10 Core Bluetooth architecture

The Bluetooth connection type capable of flexible and higher bandwidths is the asynchronous connection-less link [9, 16]. Figure 10 shows the core Bluetooth protocol layers.

The baseband layer enables the physical RF link between Bluetooth units making a connection. Link manager protocol (LMP) is responsible for link set-up between Bluetooth devices and managing security aspects such as authentication and encryption. L2CAP adapts upperlayer protocols to the baseband. It multiplexes between the various logical connections made by the upper layers. Audio data typically is routed directly to and from the baseband and does not go through L2CAP.

3.1.1 Java Standard API for Bluetooth: JSR-82

Given that our communication scheme includes a client application implemented on the information source and the server application running on a mobile phone, it is reasonable to choose a Bluetooth programming technology that is provided in nowadays mobile devices. That is why we have decided to use the standard Bluetooth Java programming API JSR-82 currently supported in a wide range of mobile devices from different manufacturers [26]. JSR-82 API allows us to establish an L2CAP point-to-point connection between client and server devices through which the neural signals are transmitted. JSR-82 supports Bluetooth standard v1.1. which is therefore the version used in our experiments.

In order to have control over the Bluetooth transmission we have programmed the client and server applications implementing the communication. In this scheme, first, an L2CAP connection is established between the master and the slave. Once communication is established, the client application running on the slave starts sending data over the connection to the master's server application. The data packet size used in the connection can be selected at compilation time and a 2 Mbyte neuronal signal of the type described in Section 2 is used as data source. The transmitter monitors the channel quality by inspecting throughput.

3.2 Transmission Over 3G

The third-generation transmission standard for mobile communication enhances GPRS (general packet radio access) in a variety of performance characteristics:

- high transmission rates up to 2 Mbps;
- high security and confidentiality;
- efficient multiple access;
- high resistance to interferences;
- global roaming;
- always on, QoS (Quality of Service);
- low cost.

In this contribution we have used the 3G technology to transmit the neural data from a mobile terminal to a remote server over public cellular networks. This remote server is an ordinary PC, a laptop, or even a remote MEA connected to a neural culture.

As the mobile device receives the neural data from the information source, they are retransmitted immediately to the remote server. Once the mobile phone is registered in the network, a profile containing all necessary parameters for the 3G transmission, such as access point, is established. The TCP, transmission control protocol, is used for the data transmission. It offers a point-to-point connection-oriented reliable link recovering a huge variety of errors dynamically and adaptively. In order to use the TCP, the transmitter (in this particular case the mobile phone) and the receiver (equivalent to the remote server in our application) shall create the terminal points of the connection, called sockets. A socket is defined by a transmission protocol (TCP is this case), an IP address, and a port number. In our experiments the mobile phone is programmed to be the client. It requests the opening of a TCP-socket to the server that is waiting for inquiries.

The application running on the mobile phone implementing both the Bluetooth and the 3G transmission is programmed in J2ME™ (due to the limited device resources). Also, the server application is programmed using Java™.

In Fig. 9 it can be observed that both the application running on the information source PC and the remote server application incorporate the JMATLink software package. This package allows the integration of MATLAB™ applications with Java™ applications.

Especially for data pre- and post-processing as well as for real-time data representation this package offers huge advantages. The data compression algorithms described in Section 2.2 are implemented in MATLAB™ and launched by JMATLink. For the evaluation of the transmission, real-time graphical data representation is required on the server, also implemented in MATLAB™ and launched by JMATLink.

3.3 Transmission Results

Due to the fact that the Bluetooth L2CAP connection is a secure channel, retransmissions assure the correct arrival of each single packet and until the acknowledgment of the former packet does not confirm its correct reception a new packet is not transmitted. For this reason, measuring transmission throughput is equivalent to measuring reception throughput. Moreover, this ensures the real-time transmission as long as the data stream-generation velocity (required transmission bandwidth as represented in Fig. 1) does not surpass the channel throughput.

In Fig. 11 the transmission mean throughput in relation to the defined packet size is represented. The mean throughput is calculated as the number of transmitted bits divided by the overall time required for transmission measured in nanoseconds. As it can be observed from Fig. 11, the mean throughput increases with the packet size. For a packet sizes smaller than 1,000 bytes the throughput is below 180 kbps. Due to the fact that the required minimum transmission data rate for neural signals, as described in Section 2 by Fig. 1 is 180 kbps, only packet sizes greater than 1,000 bytes provide real-time transmission of one neural signal. For these packet sizes

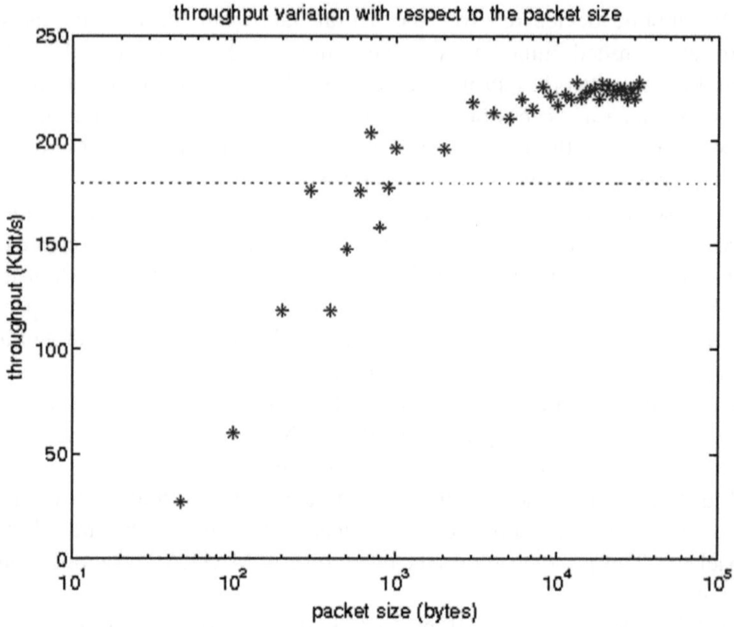

Fig. 11 Measured mean throughput with respect to the transmission packet size

(> 1,000 bytes) as the packet size increases, saturation in the mean throughput is observed. The obtained maximum mean throughput value is below 230 kbps.

Fortunately, the measured throughput values are improved by using the Bluetooth v.2. EDR (enhanced data rate). With this new standard, data rates up to 3 Mbps are achieved. Due to the limited processing and storage capabilities of the mobile phone, the maximum packet size for the Bluetooth transmission is 512 bytes. In Fig. 12 (top) the real-time evolution of the transmission throughput for a packet size of 512 bytes is represented. As it can observed, there appear peak values of up to 695.6 kbps, while the minimum value is 24.61 kbps. The mean throughput obtained for a 512 bytes packet size is of 323.1 kbps for the experiment shown in Fig. 12. Figure 12 (bottom) shows the corresponding time profile. It can be observed that throughput peak values in Fig. 12 (top) correspond to time minimum values as appears in packet nr. 8. The mean packet transmission time is calculated to be 12.67 ms. The obtained throughput allows the real-time transmission of one neural signal channel (180 kbps required for each channel). Therefore, adequate data compression before transmission is mandatory.

Analyzing the real-time throughput evolution, it is observed that less than 20% of the measured throughput values fall below the range of the mean value. Therefore, an adequate mean throughput value guaranties the channel capacity for over 80% of the time.

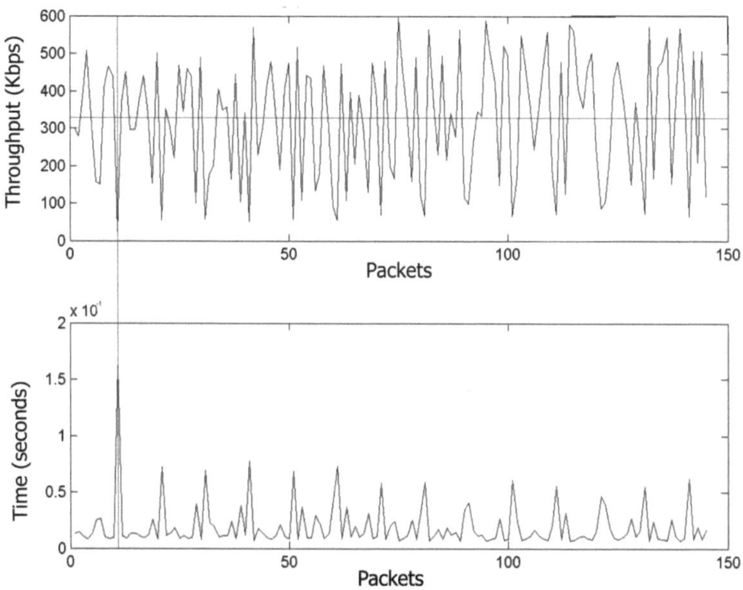

Fig. 12 Measured throughput and packet transmission time for packet size 512 bytes with EDR

The developed compression algorithm described in Section 2 is able to reduce the required data transmission rate on an average to 75% depending on the spiking activity of the particular culture.

With the transmission rate limited to 323.1 kbps and taking advantage of the compression algorithm, it is possible to transmit in real-time four neural signals on an average. In fact, the spiking activity of the neural signal is monitored in real time and the number of transmitted neural signals is adaptively recalculated. Finally, the most active channels are transmitted in real time. The measurement of the activity of a channel is performed by inspecting the actual compression rate: the more active a channel is the lower will be its compression rate.

4 Neural Signal Post-processing

At the receiver, neuronal signals are reconstructed from compressed format. For doing so, for each spike, channel number, frame number and spike time are extracted. Then, spike samples are situated in the corresponding channel and time to obtain the original spike patterns.

Besides signal reconstruction, given that electrodes record signals from multiple neurons, a classification phase is needed, where spikes are assigned to originating neurons based on a spike waveform analysis algorithm.

4.1 Decoding and Signal Reconstruction

The spike coding algorithm that allows the source signal compression is not fully reversible, given that even for excellent detection performance the ideal curve cannot be achieved. Nevertheless, as it will be shown further on, the compression algorithm takes advantage of the neural signal characteristics thus allowing signal post-processing with similar quality as without compression.

For each frame that is received by the server the decoding algorithm provides the crucial information extracted from the frame header (see Fig. 4) as input for the reconstruction algorithm. Frame by frame, the reconstruction algorithm situates, for the received channel identification number, the coded spike samples on the corresponding time scale. The resulting reconstructed signal is represented in Fig. 13. As it can be observed, the reconstructed signal (green) contains the detected spikes but does not include the noise signal when no spike is present.

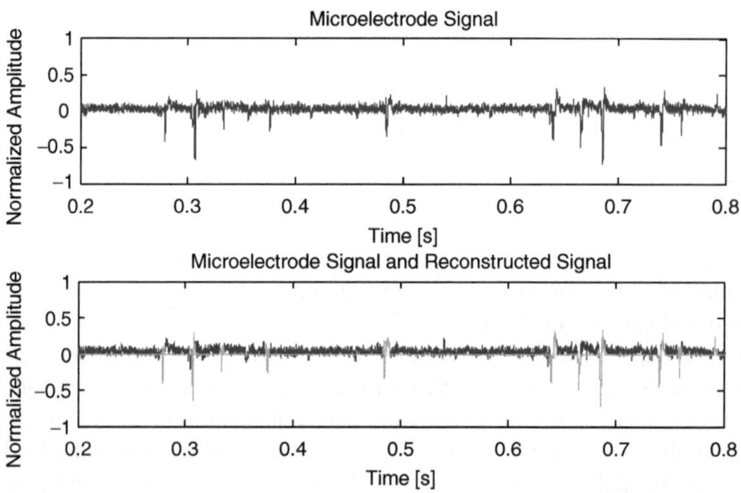

Fig. 13 Result of the reconstruction algorithm for one channel: original neural signal (*blue*) compared to the reconstructed signal (*green*)

As a consequence, from the reconstructed signal a variety of interesting parameters for the neural signal analysis can be extracted. Actually, due to the fact that not only the spike presence is transmitted but also the corresponding recorded samples, even spike classification can be tackled.

4.2 Classification

Typically, MEA are situated such that for each electrode, there exist a number of surrounding neural cells [12]. Thus, each single electrode of the MEA records the signals originating from several neural cells, obtaining multispike trains for each MEA electrode.

The exact waveform captured for each neuron depends on the neural cell itself and the geometry of the extracellular space as described in [31]. Moreover, the waveform characteristics of the captured signal are constant over time for each neuron. Exactly these different waveform characteristics can be used to identify the corresponding neuron in a single electrode recording, that is, to classify the detected multispike train. In Fig. 14 (top) the detections and spiking frequency obtained for a typical multispike train are depicted. Considering that this multispike train includes contributions of three different neural cells, detection and spiking frequency (depicted in Fig. 14 (bottom)) are computed for each of the individual neurons, thus implying prior classification. From Fig. 14 it becomes apparent that classification is a must when dealing with multispike trains. As we have seen, for each MEA electrode recording, the individual contributions of the surrounding neural cells can be distinguished using signal processing algorithms that take advantage of the similar waveform characteristics of the spikes originated by one neural cell. This implies that prior to classification a detection process shall be performed. That is the reason why compression through detection allows neural signal post-processing with similar quality parameters as those yielded for the original recorded signal. However, simultaneous firing of two or more neural cells surrounding one MEA electrode can cause overlapping of the associated waveforms deforming the resulting signal and thus, increasing the difficulty of the spike sorting task.

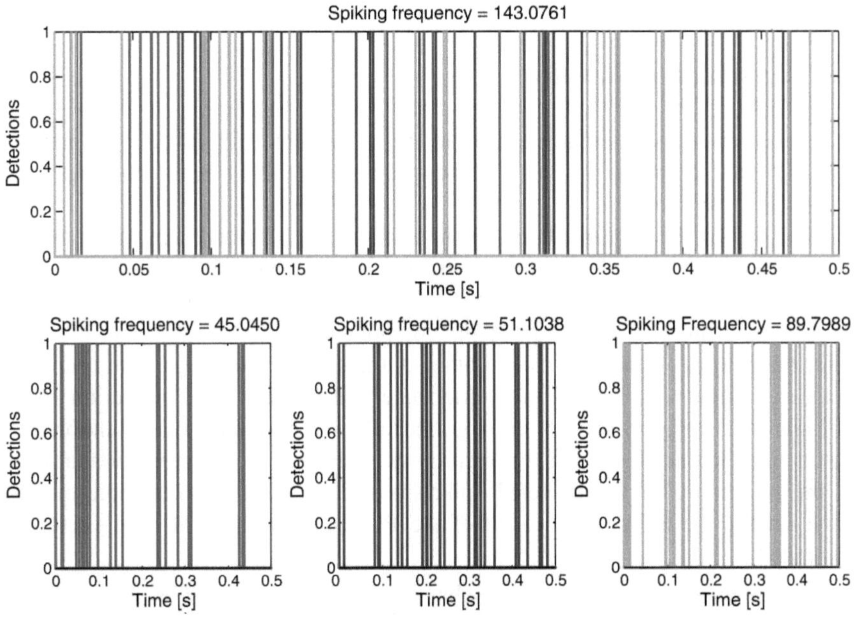

Fig. 14 The need of spike sorting in multispike trains

Any automatic classification process is based on two consecutive steps:

1. Extraction of the most relevant signal characteristics.
2. Based on the extracted characteristics, determination of the classes and the membership of each of the signals to the classes.

When it comes to the multispike train sorting problem, several characteristics extraction algorithms are suitable [12, 21]. Our choice fell on principal components analysis (PCA), selected for its demonstrated excellent performance [34]. The PCA is based on the stored reconstructed frames. Through PCA, for each of the spike sets, a set of sorted vectors that forms an orthogonal base capable of representing the spikes' subspace is obtained. These base vectors indicate the directions of maximum data variation and each spike can be represented as a scaled sum of them. Base vectors are sorted with respect to their relative contribution in representing the set of analyzed signals and, selecting the N base vectors with highest scores of this sorted list, the spikes are characterized through their projection on the selected base vectors. Exactly these projections are the extracted characteristics, principal components, used for the determination of one class, that is, the originating neural cell.

Once the characteristics (N base vectors or principal components) have been extracted, the class membership algorithm *k-means* is applied to the multispike train. This algorithm basically consists in associating to each spike the class with the closest weight center using the Euclidean distance. The weight center of the associated class is recalculated after the inclusion of the spike.

4.3 Classification Results

In order to assess performance quality of the implemented classification algorithm it is applied to both, the artificially generated signals and the real recordings from multichannel systems.

Similarly to the process described in Section 2.1, artificial signals containing two trains of spikes randomly superposed were generated and in fact, the artificial multispike train classification is used to tune interactively several parameters of the classification process such as principal component score and clustering distance. Once the optimized parameters are obtained, the classification algorithm is applied to the real signal recordings from multichannel systems described in Section 2.

Figures 15, 16, 17, and 18 show graphically the developed classification process applied to one channel of real recordings. In Fig. 15 the superposition, aligned to the minimum, of the detected spikes is shown. These spikes represent the input set for the PCA. As a result of the PCA, a sorted list of base vectors is computed.

The computed scores of the principal components of the complete input spike set that form the sorted list are shown in Fig. 16 (top). The three most scored Principal Components collect more than 81% of the overall scores, thus yielding a

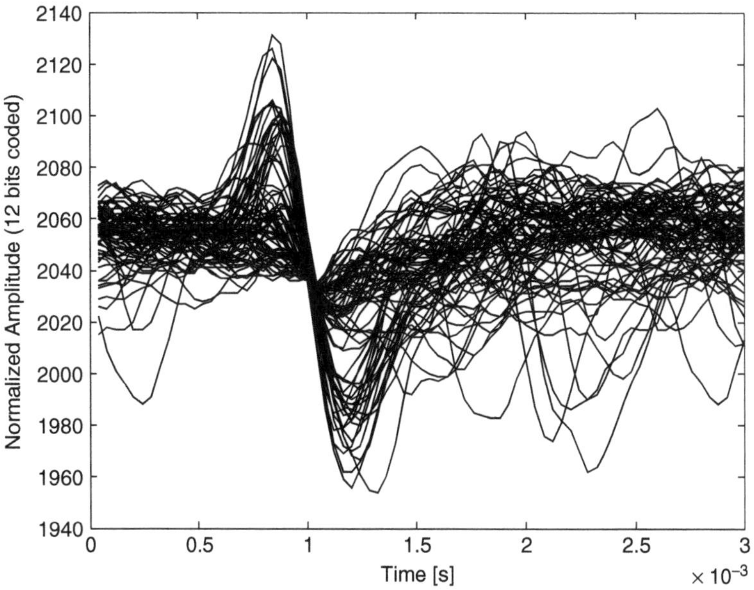

Fig. 15 Superposition of the detected spikes of one channel. Alignment is performed to the minimum value

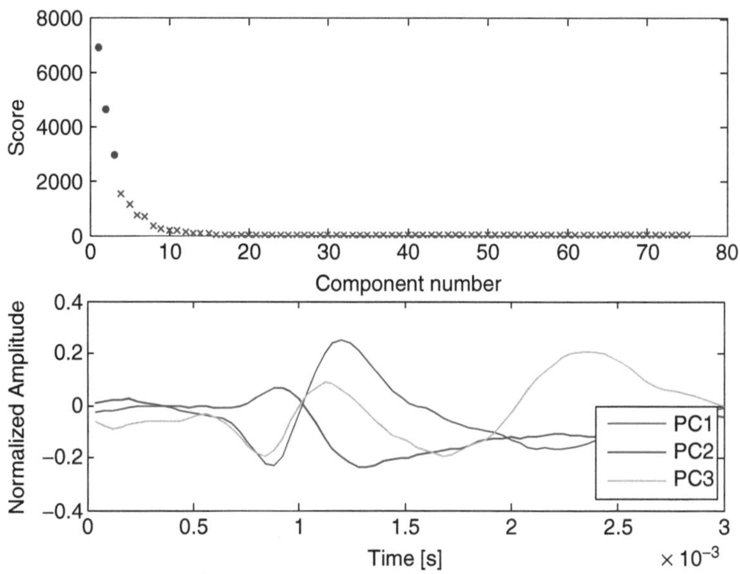

Fig. 16 Scores of principal components of the input spike set (*top*) and three most relevant principal components (*bottom*)

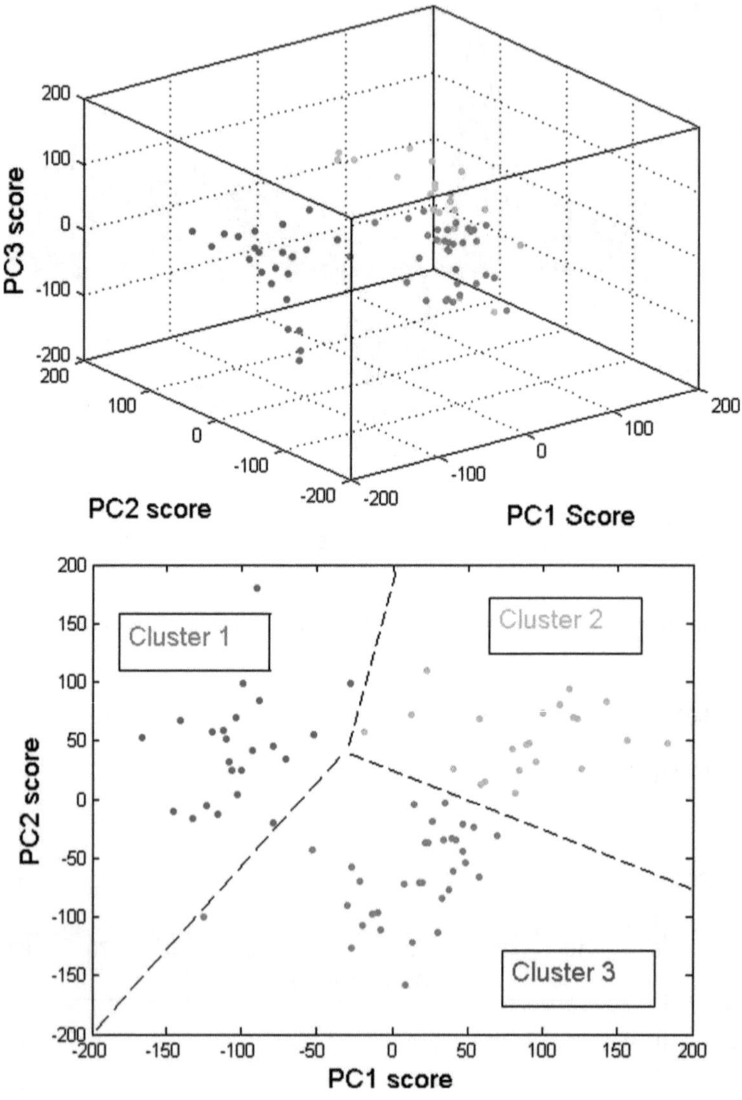

Fig. 17 3-Dimensional representation of the input spike set (*top*) and 2-dimensional projection (*bottom*)

high-fidelity representation of the complete input set when truncating at a 3-dimensional representation. Precisely these three most relevant signals are depicted in Fig. 16 (bottom). They form an orthogonal vector base that will be used for the whole input set space.

From this 3-dimensional representation the *k-means* clustering algorithm is applied to establish the membership of each spike to one determined class. Figure 17 (top) shows the 3-dimensional representation of the complete spike input

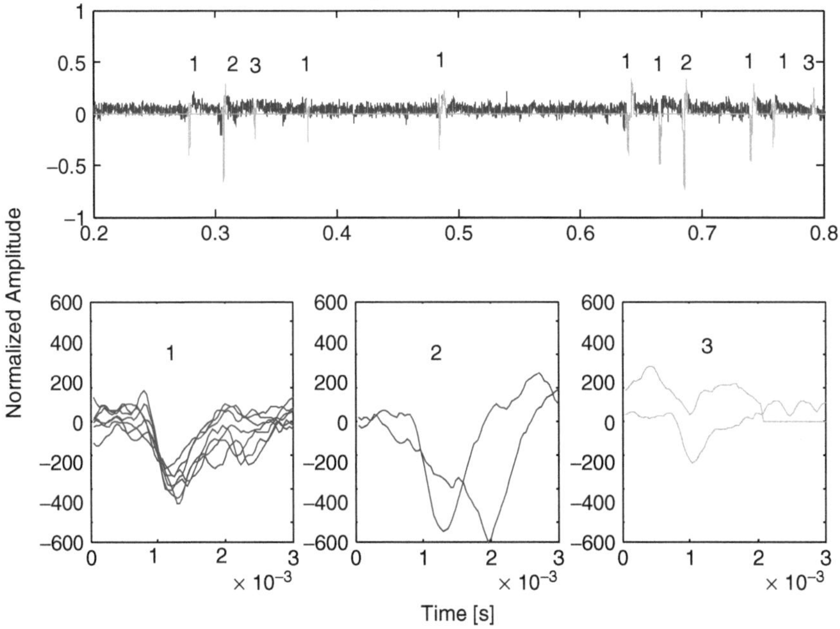

Fig. 18 Classified input spike set (*top*) and spikes belonging to class 1–3 (*bottom*)

set separating colorwise the three different classes. For the sake of clarity, the 2-dimensional projection on the two first Principal Components is also shown in Fig. 17 (bottom).

The overall result of the classification is summarized in Fig. 18 (top) that shows the input spike set separated colorwise depending on its class membership. As it can be observed from Fig. 18 (bottom), the spikes belonging to one class show similar waveforms

The classification performance for the set of artificial multispike trains yields 92% of correctly classified spikes. For the real data recorded by multichannel systems (see Section 2) the classification performance shall be assessed through evaluation of experts, similarly to the assessment of the detection quality described in Section 2.

5 Conclusions and Future Trends

In this contribution a wireless transmission system over Bluetooth and 3G is analyzed for its application to the real-time transmission of neural signals captured by implanted micro-electrode array sensors.

First, the required data rates for this type of neural signals are calculated to be not less than 180 kbps for every single micro-electrode. Thus, for an array of 64 micro-electrodes a minimum transmission rate of 1,1520 kbps is required. To be able to compress the neural signals, detection of spikes is implemented and average 75% compression rate is achieved. Detection uses a nonlinear energy operator preprocessing and automatic threshold adaptation. Results of spike detection quality assessment show successful adaptation to different input SNRs and thus eliminating the need for manual threshold setting.

Wireless transmission of these signals integrates a Bluetooth transmission from the information source to a mobile device and a data transmission from the mobile device over 3G to a remote server, without intermediate storage on the mobile phone.

The transmission rate is limited by the Bluetooth and 3G links, depending on the transmission packet size. Due to the limited resources of the mobile phone, the maximum transmission unit is also bounded, thus achieving a maximum transmission rate of 323.1 kbps. With this transmission rate, it is not possible to transmit more than one neural signal in real-time over the Bluetooth link without compression. With the developed compression algorithm the system performance is enhanced allowing real-time transmission of four neural signals considering average spiking activity.

The developed system is one step further to gain insight in brain activity for neuroscientists. Recent work indicates that there is a vast universe of possible applications [14], especially considering brain–machine interfaces (BMIs), devoted to create interfaces between the human brain and the artificial devices [11, 17, 27]. Scientists from a wide range of disciplines are working on technologies that allow patients to use brain activity signals to control mechanical or electronic devices providing technologies that allow the patients to restore lost sensory-motor functions [19, 29]. There are still fundamental questions to be solved in the neurobiology field; however, first results of brain-actuated technologies, such as neuroprostheses or neurorobots, lead to optimistic expectations.

When it comes to focus on the communication systems involved in these BMIs there is still much work to be done. The neural signal, the fundamental information source, hides significant conceptual complexity and thus, its recordings shall be made available to as many researchers as possible [8]. That is why worldwide transmission systems are so important particularly in this area. On the other hand, a fundamental requirement for any communication system aimed toward the development of clinical applications of BMIs is to be wireless in order to allow patients to wander around during neural recording and monitoring. Moreover, aspects like low power consumption (especially considering implantable devices, not only because of the body proximity but also to extent battery life) and small interference with already existing systems shall not be neglected.

Considering these requirement and the high data rates generated by implantable neural sensors, from our point of view, future applications will probably start to exploit ultra wideband (UWB) technology [24, 25, 33]. UWB presents several characteristics that make it very attractive for this particular application:

- considerable high transmission rates of over 100 Mbps
- extremely low power consumption
- no interference with other wireless technologies due to its spread transmission spectrum (short pulse transmission)

There are some additional considerations to be regarded. UWB allows only short-distance communications with these high transmission rates, which is perfectly assumable for body area networks but raises the need of a bridge between the close body field and remote stations. By now, there is no global regulation or standardization for UWB. With the announcement of the new Bluetooth standard with UWB as core transmission technology being made commercially available very soon, the solution for these issues arose. Commercial mobile phones and computers will integrate UWB technology as high data rate transmission technology and thus, mobile phones can figure as high data rate bridges between body area communications and the rest of the world.

References

1. Akin T, Najafi K, Bradley R (1998) A wireless implantable multichannel digital neural recording system for a micromachines sieve electrode. IEEE Journal of Solid State Circuits, 1, 33, 109–118
2. Anderson DJ, Oweiss KG (2003) Capturing Signal Activity and Spatial Distribution of Neurons in a Sub-Millimeter Volume. Conference Record of the 37th Asilomar Conference on Signals, Systems and Computers, 1, 387–390
3. Bluetooth special interest group: Specification of the bluetooth system (2004) http://www.bluetooth.com. Accessed 10th July 2008
4. Brown EN, Kass RE, Mitra PP (2004) Multiple neural spike train data analysis: state-of-the-art and future challenges. Nature Neuroscience, 7, 5, 456–461
5. Fernández E, Pelayo F, Romero S, Bongard M, Marín C, Alfaro A, Merabet L (2005) Development of a cortical visual neuroprosthesis for the blind: the relevance of neuroplasticity. Journal of Neural Engineering, 2, R1–R12
6. Hochberg LR, Serruya MD, Friehs GM, Mukand JA, Saleh M, Caplan AH, Branner A, Chen D, Penn RD, Donoghue JP (2006) Neuronal ensemble control of prosthetic devices by a human with tetraplegia. Nature, 442, 164–171
7. Irazoqui-Pastor P, Mody I, Judy JW (2002) Transcutaneous rf-powered neural recording device. Second Joint EMBS/BMES Conference, Houston, USA, 3, 2105–2106
8. Irazoqui-Pastor P, Mody I, Judy JW (2005) Recording brain activity wirelessly. IEEE Engineering in Medicine and Biology Magazine, 24, 6, 48–54
9. Ju MC, Park CH, Hong DK, Youn KJ, Cho JW (2002) Link management scheme of bluetooth based on channel quality estimation. Electronics Letters, 38, 15, 89–790
10. Kim K, Kim S (2000) Neural spike sorting under nearly 0 dB signal-to-noise ratio using non-linear energy operator and artificial neural-network classifier. IEEE Transactions on Biomedical Engineering, 47, 10, 1406–1411
11. Lebedev MA, Carmena JM, O'Doherty JE, Zacksenhouse M, Henriquez CS, Principe JC, Nicolelis MAL (2005) Cortical ensemble adaptation to represent velocity of an artificial actuator controlled by a brain-machine interface. The Journal of Neuroscience, 25, 19, 4681–4693
12. Letelier JC, Weber PP (2000) Spike sorting based on discrete wavelet transform coefficients. Journal of Neuroscience Methods, 2, 101, 93–106

13. Lopez-Casado C, Tejero-Calado J, Bernal-Martin A, Lopez-Gomez M, Romero-Romero M, Quesada G, Lorca J, Garcia E (2005) Network architecture for global biomedical monitoring service. 27th Annual International Conference of the Engineering in Medicine and Biology Society, Shanghai, China, pp. 2433–2436
14. Martinoiaa S, Sanguinetib V, Cozzib L, Berdondinic L, van Peltd J, Tomase J, Le Massonf G, Davideg F (2004) Towards an embodied in vitro electrophysiology: the NeuroBIT project. Neurocomputing, 60, 58, 1065–1072
15. Mojarradi M, Binkley D, Blalock B, Anderson R, Ulshoefer N, Johnson T (2003) A miniaturized neuroprosthesis suitable for implantation into the brain. IEEE Transactions on Neural Systems and Rehabilitation Engineering, 3, 11, 38–42
16. Morrow R, (2000) Connecting with a bluetooth piconet. Fall Wireless Symposium/Portable By Design Conference and Exhibition, Chicago
17. Nicolelis MAL (2001) Actions from thoughts. Nature, 409, 403–407
18. Obeid I, Nicolelis MAL, Wolf PD (2004a) A multichannel telemetry system for single unit neural recording. The Journal of Neuroscience Methods, 133, 33–38
19. Obeid I, Nicolelis MAL, Wolf PD (2004b) A low power multichannel analog front end for portable neural signal recordings. The Journal of Neuroscience Methods, 133, 27–32
20. Obeid I, Wolf PD (2004) Evaluation of Spike-detection algorithms for a brain-machine interface application. IEEE Transactions on Biomedical Engineering, 51, 6, 905–911
21. Olson BP, Si J, Hu J, He J (2005) Closed-loop cortical control of direction using support vector machines. IEEE Transactions on Neural systems and Rehabilitation Engineering, 13, 1, 72–80
22. Ryugo DK, Kretzmer EA, Niparko JK. (2005) Restoration of auditory nerve synapses in cats by cochlear implants. Science, 310, 5753, 1490–1492
23. Salamon D, Bei A, Grigioni M, Gianni M, Liberti M, D'Inzeo G, Luca SD (2005) Indoor telemedicine in hospital: a pda-based flexible solution for wireless monitoring and database integration. 27th Annual International Conference of the Engineering in Medicine and Biology Society, Shanghai, China, pp. 386–389
24. Tarín C, Martí P, Traver L, Cardona N, Díaz JA, Antonino E (2007) UWB Channel Measurements for hand-portable devices: a comparative study. IEEE International Symposium on Personal, Indoor and Mobile Radio Communications, Athens, Greece, pp. 1–5
25. Tarín C, Traver L, Martí P, Cardona N, Díaz JA, Cabedo M (2007) UWB Channel measures for hand-portable and wearable devices. IEEE International Conference on Wireless and Mobile Computing, Networking and Communications, New York, USA, pp. 1–6
26. Tarín C, Traver L, Santamaría JF, Martí P, Cardona N (2007) Bluetooth-3G wireless transmission system for neural signal telemetry. IEEE Wireless Telecommunications Symposium, Pomona, California, USA, pp. 1–6
27. Taylor DM, Helms Tillery SI, Schwartz AB (2002) Direct cortical control of 3D neuroprosthetic devices. Science, 296, 1892–1832
28. Universite Paris Descartes, UFR Biomedicale. http://www.biomedicale.univ-paris5.fr/SpikeOMatic/Data.html. Accessed 10th July 2008
29. Wessberg J, Stambaugh CR, Kralik JD, Beck PD, Laubach M, Chapin JK (2000) Real-time prediction of hand trajectory by ensembles of cortical neurons in primates. Nature, 408, 6810, 361–365
30. Wise KD, Anderson DJ, Hetke JF, Kipke DR, Njafi K (2004) Wireless implantable microsystems: High-density electronic interfaces to the nervous system. Proceedings of the IEEE, 92,1, 76–97
31. Wood F, Black MJ, Vargas-Irwin C, Fellows M, Donogue JP (2004) On the variability of manual spike sorting. IEEE Transactions on Biomedical Engineering, 51, 6, 912–918
32. Yu SN, Cheng JC (2005) A wireless physiological signal monitoring system with integrated bluetooth and wifi technologies. 27th Annual International Conference of the Engineering in Medicine and Biology Society, Shanghai, China, pp. 2203–2206

33. Zasowski T, Althaus F, Stäger M, Wittneben A, Tröster G (2003) UWB for noninvasive wireless body area networks: Channel measurements and results. IEEE Conference on Ultra Wideband Systems and Technologies, pp. 1–10
34. Zumsteg ZS, Kemere C, O'Driscoll S, Santhanam G, Ahmed RE, Shenoy KV, Meng TH (2005) Power feasibility of implantable digital spike sorting circuits for neural prosthetic systems. IEEE Transactions on Neural Systems and Rehabilitation Engineering, 13, 3, 272–279

The Modified Max-Log-MAP Turbo Decoding Algorithm by Extrinsic Information Scaling for Wireless Applications

Mustafa Taskaldiran, Richard C.S. Morling, and Izzet Kale

Abstract Turbo codes have found use in various wireless communication applications and have been incorporated into important standards like 3GPP and DVB. The iterative nature of turbo decoding algorithms increases their complexity compare to the conventional FEC decoding algorithms. A simple but effective technique to improve the performance of the Max-Log-MAP turbo decoding algorithm is to scale the extrinsic information exchanged between two MAP decoders. A comprehensive analysis of the selection of the scaling factors according to channel conditions and decoding iterations is presented in this chapter. Choosing a constant scaling factor for all SNRs and iterations is compared with the best scaling factor selection for changing channel conditions and decoding iterations. It is observed that a constant scaling factor for all channel conditions and decoding iterations is the best solution and provides a 0.2–0.4 dB gain over the standard Max-Log-MAP algorithm. Therefore, a constant scaling factor should be chosen for the best compromise.

1 Introduction

In his 1948 paper [1], Shannon defined the limits of a communication system. He proved that there exists error-correcting codes which can provide arbitrarily high reliability of transmission for information rates below the channel capacity. In spite of all efforts to find such error control codes, the gap between the Shannon limit and the practice was still 2 dB until 1993. A major advancement in the channel coding area was introduced by Berrou et al. in 1993 by the advent of turbo codes [2]. Turbo codes have shown the best forward error correction (FEC) performance known up to now. Turbo codes are revolutionary in the sense that they allow reliable data

---

M. Taskaldiran (✉)
Applied DSP and VLSI Research Group, Department of Electronic Systems,
University of Westminster, London, W1W 6UW, United Kingdom
e-mail: m.taskaldiran@wmin.ac.uk

S. Powell, J.P. Shim (eds.), *Wireless Technology*, Lecture Notes in Electrical Engineering 44, DOI 10.1007/978-0-387-71787-6_13,

transmission within a half decibel of the Shannon Limit. At first, the extraordinary performance of turbo codes encountered some doubts by the communication community. However, their performance has been verified by many researchers in a short time after the emergence of turbo codes. A massive amount of research effort has been performed to facilitate the energy efficiency of turbo codes [3]. The superior performance of turbo codes has been studied and well understood [4]. As a result, turbo codes have been incorporated into many standards used by the NASA Consultative Committee for Space Data Systems (CCSDS), Digital Video Broadcasting (DVB), both Third Generation Partnership Project (3GPP) standards for IMT-2000, and Wideband CDMA which requires throughputs from 2 Mb/s to several 100 Mb/s.

The iterative nature of turbo decoding algorithms increases their complexity compare to conventional FEC decoding algorithms. Two iterative decoding algorithms, soft-output-viterbi algorithm (SOVA) and Maximum A posteriori Probability (MAP) Algorithm require complex decoding operations over several iteration cycles. So, for real-time implementation of turbo codes, reducing the decoder complexity while preserving bit-error-rate (BER) performance is an important design consideration.

In this chapter, a modification to the Max-Log-MAP algorithm is presented. This modification is to scale the extrinsic information exchange between the constituent decoders. The remainder of this chapter is organized as follows: An overview of the turbo encoding and decoding processes, the MAP algorithm, and its simplified versions the Log-MAP and Max-Log-MAP algorithms are presented in Section 1. The extrinsic information scaling is introduced, simulation results are presented, and the performance of different methods to choose the best scaling factor is discussed in Section 2. Section 3 discusses trends and applications of turbo coding from the perspective of wireless applications.

1.1 Turbo Encoder

A generic structure for turbo encoding based on parallel concatenation of two Recursive Systematic Convolutional (RSC) encoders is given in Fig. 1. Two identical RSC encoders produce the redundant data as parity bits. The input data stream and parity bits are combined in series to form the turbo coded word. The size of the input data word may vary from 40 to 5114 bits for UMTS [5] and take specified values such as 378, 570, and 20730 for CDMA2000 [6] turbo coding which are the two main standards of 3GPP and 3GPP2, respectively.

The interleaver is the crucial part of turbo encoding as it shapes the weight distribution of the code in a way to produce low-weight code words. Opposite to their non-recursive counterparts, RCS encoders can only be terminated by certain terminating data sequences. The interleaver separating two RCS encoders prevents at least one of the encoders to terminate quickly. It is obvious that a data sequence terminating after a long period has a large Hamming distance and hence provides

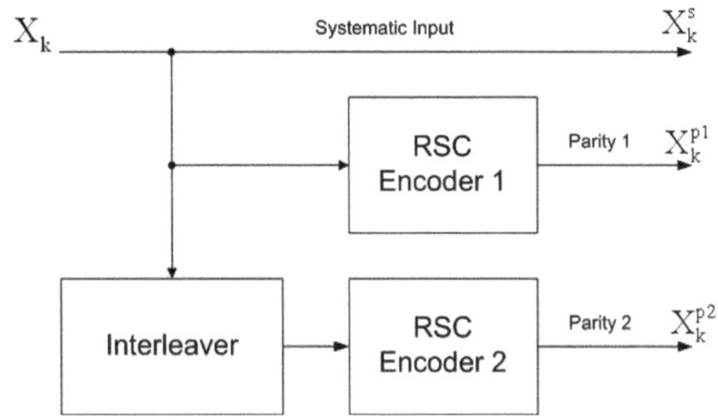

Fig. 1 Generic turbo encoder

better error protection [3]. This improvement is called the *interleaver gain* which is one of the main reasons of the excellent performance of turbo codes [7].

The interleaver design also affects the turbo decoder performance by reducing the degree of correlation between the soft-output of each decoder which becomes the extrinsic information to the other decoder (decoder1 and decoder2 in Fig. 2). As the degree of correlation between these two soft-information decreases, the performance of the turbo decoder increases [8].

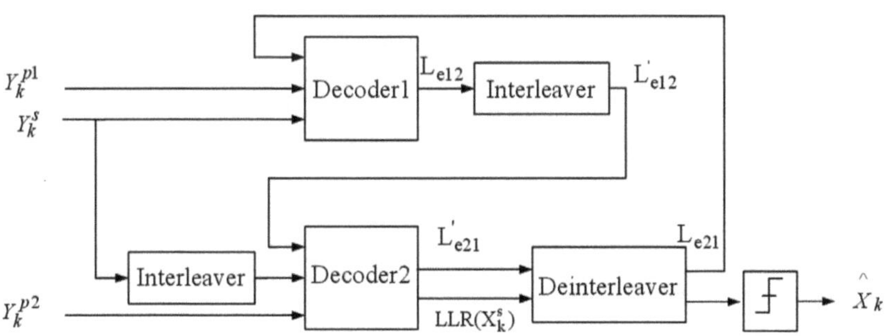

Fig. 2 Iterative turbo decoding

1.2 Turbo Decoder

In a typical turbo decoding system (see Fig. 2), two decoders operate iteratively and pass their decisions to each other after each iteration. These decoders should produce soft-outputs to improve the decoding performance. Such a decoder is called a soft-input soft-output (SISO) decoder [9]. Each decoder operates not only on its own input but also on the other decoder's incompletely decoded output which resembles the operation principle of turbo engines. This analogy between the operation of the turbo decoder and the turbo engine gives this coding technique its name, "turbo codes" [10].

Turbo decoding process can be explained as follows: Encoded information sequence X_k is transmitted over an additive white gaussian noise (AWGN) channel, and a noisy received sequence Y_k is obtained. Each decoder calculates the log-likelihood ratio (LLR) for the kth data bit d_k, as

$$L(d_k) = \log\left[\frac{P(d_k = 1|Y)}{P(d_k = 0|Y)}\right] \tag{1}$$

LLR can be decomposed into three independent terms, as

$$L(d_k) = L_{apri}(d_k) + L_c(d_k) + L_e(d_k) \tag{2}$$

where $L_{apri}(d_k)$ is the a priori information of d_k, $L_c(x_k)$ is the channel measurement, and $L_e(d_k)$ is the extrinsic information exchanged between the constituent decoders. Extrinsic information from one decoder becomes the a priori information for the other decoder at the next decoding stage. L_{e12} and L_{e21} in Fig. 1 represent the extrinsic information from decoder1 to decoder2 and decoder2 to decoder1, respectively.

LLR computations can be performed by using one of the two main turbo decoding algorithms SOVA and MAP algorithms. The MAP algorithm seeks for the most likely data sequence, whereas SOVA, which is a modified version of the Viterbi algorithm, seeks for the most likely connected path through the encoder trellis. The MAP algorithm is a more complex algorithm compared to SOVA. At high SNR, the performance of SOVA and MAP are almost the same. However, at low signal-to-noise ratios (SNRs) MAP algorithm is superior to SOVA by 0.5 dB or more [9]. The following sections explain the MAP algorithm and its simplified versions Log-MAP and Max-Log-MAP algorithms.

1.3 The MAP Algorithm

The MAP algorithm is an optimal but computationally complex SISO algorithm. The Log-MAP and Max-Log-MAP algorithms are simplified versions of the MAP algorithm.

MAP algorithm calculates LLRs for each information bit as

$$L(d_k) = \ln\left[\frac{\sum\limits_{S_k}\sum\limits_{S_{k-1}} \gamma_1(S_{k-1},S_k)\alpha(S_{k-1})\beta(S_k)}{\sum\limits_{S_k}\sum\limits_{S_{k-1}} \gamma_0(S_{k-1},S_k)\alpha(S_{k-1})\beta(S_k)}\right] \tag{3}$$

where α is the forward state metric, β is the backward state metric, γ is the branch metric, and S_k is the trellis state at trellis time k. Forward state metrics are calculated by a forward recursion from trellis time $k = 1$ to, $k = N$ where N is the number of information bits in one data frame. Recursive calculation of forward state metrics is performed as

$$\alpha_k(S_k) = \sum_{j=0}^{1} \alpha_{k-1}(S_{k-1})\gamma_j(S_{k-1},S_k) \tag{4}$$

Similarly, the backward state metrics are calculated by a backward recursion from trellis time $k = N$ to, $k = 1$ as

$$\beta_k(S_k) = \sum_{j=0}^{1} \beta_{k+1}(S_{k+1})\gamma_j(S_k,S_{k+1}) \tag{5}$$

Branch metrics are calculated for each possible trellis transition as

$$\gamma_i(S_{k-1},S_k) = A_k P(S_k|S_{k-1}) \exp\left[\frac{2}{N_o}\left(y_k^s x_k^s(i) + y_k^p x_k^p(i,S_{k-1},S_k)\right)\right] \tag{6}$$

where $i = (0,1)$, A_k is a constant, x_k^s and x_k^p are the encoded systematic data bit and parity bit, and, y_k^s and y_k^p are the received noisy systematic data bit and parity bit, respectively.

1.4 The Log-MAP Algorithm

To avoid complex mathematical calculations of MAP decoding, computations can be performed in the logarithmic domain. Furthermore, logarithm and exponential computations can be eliminated by the following approximation

$$\max{}^*(\mathrm{x,y}) \triangleq \ln\left(\mathrm{e^x} + \mathrm{e^y}\right) = \max(\mathrm{x,y}) + \log(1 + \mathrm{e}^{-|\mathrm{y}-\mathrm{x}|}) \tag{7}$$

The last term in max*(.) operation can easily be calculated by using a look-up table (LUT).

So Equations (3), (4) and (5), (6) become

$$\begin{aligned} L(d_k) = {} & \max_{(S_{k-1},S_k,1)}{}^* \left(\bar{\gamma}_1(S_{k-1},S_k) + \bar{\alpha}_{k-1}(S_{k-1}) + \bar{\beta}_k(S_k)\right) \\ & - \max_{(S_{k-1},S_k,0)}{}^* \left(\bar{\gamma}_0(S_{k-1},S_k) + \bar{\alpha}_{k-1}(S_{k-1}) + \bar{\beta}_k(S_k)\right) \end{aligned} \tag{8}$$

$$\bar{\alpha}_k(S_k) = \max_{(S_{k-1},i)}{}^* \left(\bar{\alpha}_{k-1}(S_{k-1}) + \bar{\gamma}_i(S_{k-1},S_k)\right) \tag{9}$$

$$\bar{\beta}_k(S_k) = \max_{(S_k,i)}{}^* \left(\bar{\beta}_{k+1}(S_{k+1}) + \bar{\gamma}_i(S_k,S_{k+1})\right) \tag{10}$$

$$\bar{\gamma}_i(S_{k-1},S_k) = \frac{2}{N_o}\left(y_k^s x_k^s(i) + y_k^p x_k^p(i,S_{k-1},S_k)\right) + \ln\left(P(S_k|S_{k-1})\right) + K \tag{11}$$

where K is a constant.

1.5 The Max-Log-MAP Algorithm

The correction function $f_c = \log(1 + e^{-|y-x|})$ in the max* (.) operation can be implemented in different ways. The Max-log-MAP algorithm simply neglects the correction term and approximates the max* (.) operator as

$$\ln(e^x + e^y) \approx max(x,y) \tag{12}$$

at the expense of some performance degradation.

This simplification eliminates the need for an LUT required to find the corresponding correction factor in the max* (.) operation. The performance degradation due to this simplification is about 0.5 dB compared to the Log-MAP algorithm [11].

2 Extrinsic Information Scaling

The extrinsic information exchanged between the constituent decoders can be scaled to improve the performance of turbo decoding [12–14]. With this modification Equation (11) for branch metric calculations can be rewritten as

$$\bar{\gamma}_i(S_{k-1},S_k) = \frac{2}{N_o}\left(y_k^s x_k^s(i) + y_k^p x_k^p(i,S_{k-1},S_k)\right) + s_d \ln\left(P(S_k|S_{k-1})\right) + K \tag{13}$$

The only modification is the scaling factor s_d where $d = 1,2$ for decoder1 and decoder2, respectively.

Extrinsic information scaling has been proposed to compensate for the optimistic LLR calculations of SOVA [14]. A gain of 0.4 dB has been reported for a code of memory length 4 at BER of 10^{-4} [14]. Scaling factor modification has also been applied and tested on the Max-Log-Map algorithm. Authors of [12] have reported 0.2–0.4 dB gain over the standard algorithm for 3GPP standards. They used a constant scaling factor of 0.7. In [13], scaling factor optimization for Max-Log-Map decoding is explained as mutual information combining which is the evolution of the information exchange between the two MAP decoders. The best scaling factors for each iteration were calculated for different SNRs by off-line computation. The performance difference between the modified Max-Log-Map and Log-Map was reported as 0.05 dB for UMTS-based turbo coding [13]. The performance improvement introduced by the scaling factor modification is explained as the correction of the accumulated bias due to maximum (max) operation in the Max-Log-Map algorithm [13].

2.1 Simulation Results

A constant scaling factor over all SNRs and decoding iterations improves the Max-Log-MAP decoding significantly. The best scaling factors for different SNRs and decoding iterations are found by off-line computations. These scaling factors are obtained via simulations by choosing the scaling factors corresponding to the minimum BER.

A turbo code of rate $R = 1/3$, memory length $m = 3$, generator polynomial $(13, 15)_{oct}$ is simulated to obtain the best scaling factors for different SNRs and decoding iterations. Table 1 shows the best scaling factors for iterations 1–6 and SNR values of 0–1.5 dB. Table 2 shows the best scaling factors after six iterations only for different SNRs assuming a constant scaling factor for both decoders. The performance of the modified algorithm is compared with the standard Max-Log-Map and Log-Map algorithms. Figure 3 and 4 show the BER performances of the Log-Map, the Max-Log-Map, and the modified Max-Log-Map with scaling factor 0.7 after 6 decoding iterations for interleaver lengths 5114 and 1024, respectively. A constant scaling factor (0.7) provides approximately 0.4 dB improvement over the standard Max-Log-Map algorithm at a BER of 10^{-4}. Table 3 shows BER values at E_b/N_o=1 dB, for an interleaver length of 1024.

From the simulation results, it is observed that changing scaling factors for different SNRs/iterations or just for SNRs does not improve the decoding performance. As it is shown in Table 3, BER values for three different methods of scaling factor modification are almost identical. Although, changing scaling factors for SNRs (and decoding iterations) provides an improvement over the standard Max-Log-Map algorithm, this improvement is observed to be equal to the improvement obtained by a constant choice of the scaling factor. The performance difference between the Log-Map and the modified Max-Log-Map is around 0.1 dB as observed from simulations.

Table 1 Scaling factors for different SNRs and iterations for decoder1(D1) and decoder2 (D2) (R = 1/3, interleaver length = 1024, generator polynomial $(13, 15)_{oct}$)

Iterations	1		2		3		4		5		6	
E_b/N_o	D1	D2	D1	D2	D1	D2	D1	D2	D1	D2	D1	D2
0	0*	0.5	0.5	0.6	0.6	0.6	0.6	0.6	0.6	0.6	0.6	0.7
0.25	0	0.6	0.6	0.6	0.6	0.6	0.6	0.7	0.6	0.7	0.7	0.7
0.50	0	0.6	0.6	0.7	0.7	0.7	0.7	0.7	0.7	0.7	0.8	0.7
0.75	0	0.6	0.7	0.7	0.7	0.7	0.7	0.8	0.8	0.7	0.7	0.8
1	0	0.6	0.7	0.8	0.8	0.7	0.8	0.6	0.5	0.7	0.4	0.8
1.25	0	0.7	0.9	0.7	0.5	0.9	0.5	0.8	0.4	0.8	0.3	1
1.5	0	0.7	0.7	0.8	0.5	1	0.5	0.9	0.4	0.9	0.4	0.8

*No extrinsic information from decoder2 to decoder1 for the first iteration.

Table 2 Scaling factors for different SNRs (R = 1/3, interleaver length = 5114, generator polynomial $(13, 15)_{oct}$)

E_b/N_o (dB)	0	0.25	0.5	0.75	1	1.25	1.5
Dec1	0.6	0.6	0.7	0.7	0.9	0.5	0.4
Dec2	0.6	0.7	0.7	0.8	0.8	0.8	0.8

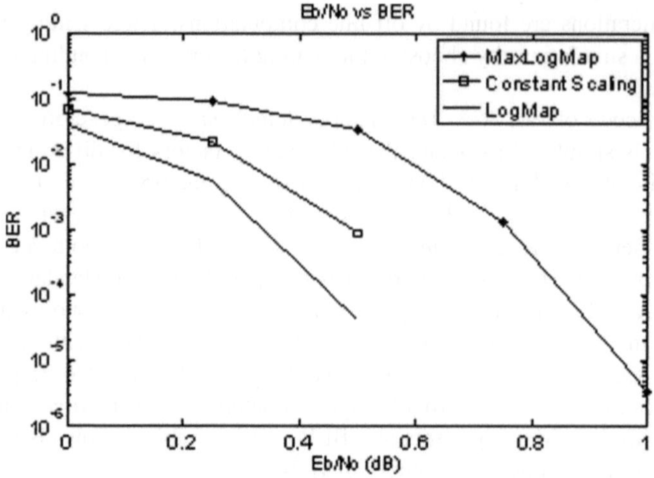

Fig. 3 BER versus E_b/N_o for Log-MAP, Max-Log-MAP, and modified Max-Log-MAP with scaling factor = 0.7 (interleaver length 5114, six iterations)

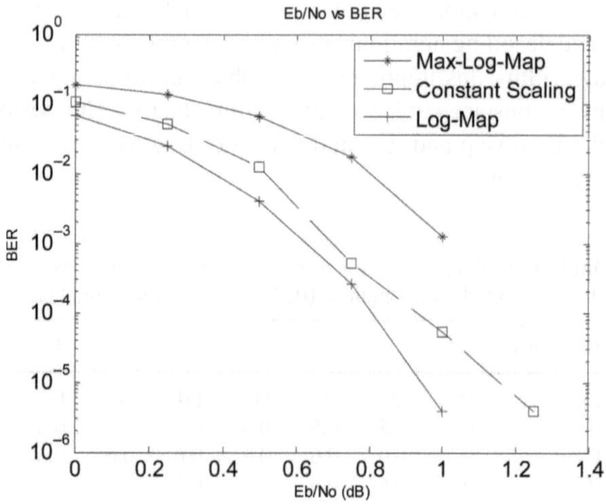

Fig. 4 BER versus E_b/N_o for Log-MAP, Max-Log-MAP, and modified Max-Log-MAP with scaling factor = 0.7 (interleaver length 1024, six iterations)

3 Trends and Applications of Turbo Codes

Near Shannon-capacity performance of turbo codes have attracted many researchers to investigate the principles in which turbo coding is based on and to apply these principles for development of new codes. Turbo coding and iterative decoding principles have found practical applications just after their discovery [15].

An important class of binary linear block code, Low-Density Parity Check (LDPC) codes [16], or Gallager codes were originally invented in 1963 and found practical applications only after the emergence of turbo decoding principles.

Table 3 BER values for each iteration at E_b/N_o = 1dB (R = 1/3, interleaver length = 1024, generator polynomial $(13, 15)_{oct}$)

	BER				
Iteration	Log-Map	Max- Log-Map	Constant Sf*	Sf for SNRs	Sf for SNRs and iterations
1	0.0396	0.0509	0.0467	0.0482	0.0462
2	0.0125	0.0301	0.0186	0.0239	0.0180
3	0.0034	0.0188	0.0067	0.0119	0.0064
4	0.0012	0.0132	0.0028	0.0068	0.0028
5	0.0006	0.0101	0.0016	0.0044	0.0016
6	0.0004	0.0087	0.0011	0.0034	0.0010

(*Sf: Scaling Factor)

Turbo-like codes have been proposed to reduce the complexity of iterative turbo decoding while providing near Shannon-capacity performance [17]. Repeat Accumulate (RA) [18], Irregular Repeat Accumulate (IRA) [19], and Accumulate Repeat Accumulate (ARA) [20] codes are some of the recently invented turbo-like codes.

Turbo codes have been incorporated into many important wireless protocols from deep space communications to mobile communications. They are used extensively in 3G mobile telephony standards. MediaFLO, terrestrial mobile television system from Qualcomm has integrated turbo codes as the error-correction algorithm. NASA missions such as Mars Reconnaissance Orbiter use turbo codes, as an alternative to Reed Solomon-Viterbi codes. The European Space Agency′s first mission to the moon SMART-1, launched in September 2003, adopted turbo codes. IEEE 802.16, a wireless metropolitan network standard, also uses turbo coding. DVB-RCS, DVB-RCT, INMARSAT, EUTELSAT, and BRAN are other systems which use turbo codes as the channel coding algorithm.

Improving the coding gain obtained by channel coding enables communication systems to work at a lower power consumption and also lower cost while preserving the quality of communication. For deep space applications, 1 dB coding gain would save millions of dollars by reducing size and weight of the required antenna and equipment (like solar panels). For mobile cellular systems, power consumption of mobile handsets can be reduced and also the quality of service can be increased by increasing the possible number of users in the cell.

As turbo codes have already been adopted in important wireless communication standards like 3GPP and DVB-RCS, there are still research areas worth to investigate aiming to reduce the complexity, power dissipation, and throughput of turbo decoders.

The most obvious of these is to improve the memory access of the turbo decoder. It has been shown that most of the energy in iterative decoding is consumed during memory access. Up to 70% of power saving, compared to conventional implementation, is possible by reducing the memory requirement [21]. The interleaver stage introduced at the encoder side complicates the memory addressing. New interleaver designs providing an easier memory access during iterative decoding have been proposed [15].

Longer interleaver lengths provide better error-protection. However, the latency and complexity introduced by the interleaver prevents the use of long interleaver lengths in real-time wireless applications. In addition, a complex algorithm is used to define the interleaver addresses of the turbo code interleaver in the 3GPP standard [5]. The easiest approach is to generate the addresses for a certain frame length and save them into a ROM. The maximum block size in 3GPP UMTS is 5114 which can be addressed by 13 bits. Instead of saving all address mappings into a ROM, a better approach to find interleaver addresses is to develop an interleaver address generator which is generic for all frame lengths [22].

Turbo codes have not only provided an excellent near Shannon-capacity performance but also have inspired the development of new codes and rediscovery of some other high-performance codes untouched for a long time, like LDPC codes. The massive research effort performed just after the invention of turbo codes and their rapid adaptation into many important communication standards show that the research and development in turbo codes and turbo-like codes will continue in the near future.

4 Conclusions

In this chapter, principles of turbo coding and its applications in wireless communications have been discussed. The emphasis has been given on an algorithm modification to improve the BER performance of the Max-Log-MAP algorithm which is the reduced complexity version of the Log-MAP algorithm.

The performance gap between the Max-Log-MAP and the Log-MAP algorithms can be compensated by scaling the extrinsic information exchange between two constituent MAP decoders. This modification in the Max-Log-MAP algorithm can be implemented simply by multiplying the extrinsic information by a scaling factor. The modified Max-Log-MAP algorithm is simulated by choosing this scaling factor as a constant as well as choosing the best scaling factors for different SNRs and decoding iterations. Simulation results show that there is almost no performance gain when we adaptively change the scaling factor with different channel conditions and for different decoding iterations against keeping the scaling factor constant. On the other hand, a proper choice of the scaling factor provides 0.4 dB improvement for the Max-Log-MAP algorithm. This optimum constant scaling factor is found to be 0.7 from our simulations.

References

1. C.E. Shannon, A mathematical theory of communication. Bell System Technical Journal, 1948. 27: 379–423, 623–656.
2. C. Berrou, A. Glavieux, and P. Thitimajshima, Near Shannon limit error-correcting coding and decoding: Turbo-codes. In Proc. ICC′93, 1993. 2: 1064–1070.
3. A. Burr, Turbo-codes: the ultimate error control codes? Electronics & Communication Engineering Journal, 2001. 13(4): 155–165.

4. S. Benedetto and G. Montorsi, Unveiling turbo codes: some results on parallel concatenated coding schemes. IEEE Transactions on Information Theory, 1996. 42(2): 409–428.
5. European Telecommunication Standards Institute, Universal Mobile Telecommunications System (UMTS): Multiplexing and channel coding (TDD),3GPP TS 25.222 version 7.3.0 Release 7, p. 18–23, May 2005.
6. Third Generation Partnership Project 2, Physical Layer Standard for cdma2000 Spread Spectrum Systems, C.S0002-D, version 2.0, p. 2.97–2.105, Sept.,2005.
7. M.C. Valenti and J. Sun, Turbo Codes. Handbook of Rf and Wireless Technologies, ed. F. Dowla. 2004, London: NewNesPress. pp. 375–399.
8. H.R. Sadjadpour, N.J.A. Sloane, and G. Nebe, Interleaver design for turbo codes. IEEE Journal On Selected Areas In Communications, 2001. 19(5): 831–837.
9. B. Sklar, Digital Communications: Fundamentals and Applications. Second ed. Fundamentals of Turbo Codes. 2001, NJ: Prentice Hall.
10. C. Heegard and S.B. Wicker, Turbo Coding. 1 ed. 1999, Boston: Kluwer Academic Publisher.
11. M.C. Valenti and J. Sun, The UMTS turbo code and an efficient decoder implementation suitable for software-defined radios. International Journal of Wireless Information Networks, 2001. 8(4): 203–214.
12. J. Vogt and A. Finger, Improving the Max-Log-MAP turbo decoder. Electronics Letters, 2000. 23: 1937–1939.
13. H. Claussen, H.R. Karimi, and B. Mulgrew, Improved max-log map turbo decoding using maximum mutual information combining. 14th IEEE Proceedings on Personal, Indoor and Mobile Radio Communications, 2003. 1: 424–428.
14. L. Papke, P. Robertson, and E. Villebrun, Improved decoding with the SOVA in a parallel concatenated (Turbo-code) scheme. IEEE International Conference on Communications, 1996. 1: 102–106.
15. B. Vucetic et al., Recent advances in turbo code design and theory. Proceedings of the IEEE Custom Integrated Circuits Conference (CICC), 2007. 95(6): 1323–1344.
16. R.G. Gallager, Low-Density Parity-Check Codes. Cambridge, MA: MIT Press, 1963.
17. A. Abbasfar, D. Divsalar, and K. Yao, A class of turbo-like codes with efficient and practical high-speed decoders. IEEE Military Communications Conference, MILCOM 2004, 2004. 1: p. 245–250.
18. D. Divsalar, H. Jin, and R. McEliece, "Coding theorems for turbo-like codes," in Proc. Allerton Conf., 1998. 201–210.
19. H. Jin, A. Khandekar, and R. McEliece, "Irregular repeat-accumulate codes," in Proc. 2nd Int. Symp. Turbo Codes, 2000. 1–8.
20. A. Abbasfar, D. Divsalar, and K. Yao, Accumulate-Repeat-accumulate codes. IEEE Transactions on Communications, 2007. 55: 692–702.
21. G. Masera et al., VLSI architectures for turbo codes. IEEE Transactions on Very Large Scale Integration (VLSI) Systems, 1999. 7(3): 369–379.
22. I. Ahmed and T. Arslan, A low energy VLSI design of random block interleaver for 3GPP turbo decoding. 2006 IEEE International Symposium on Circuits and Systems, 2006. ISCAS 2006. Proceedings, 2006. 4.

Getting Network Simulation Basics Right – A Note on Seed Setting Effects for the ns-2 Random Number Generator

Martina Umlauft and Peter Reichl

Abstract The ns-2 network simulator is one of the most widely used packet network simulators. Since version 2.1b9, it uses the MRG32k3a random number generator (RNG) proposed by L'Ecuyer, replacing the previous minimal standard multiplicative linear congruential generator by Park and Miller to remedy the problems of sensitivity to seeds and short-period length. Unfortunately, due to bad documentation and re-use of old scripts many people still wrongly use the old API functions to explicitly set seeds. While the old RNG required this, in the current MRG32k3 implementation the same approach leads to overriding the automatic seed generation of the new generator which can result in correlation between the generated random values. Using a wired and a wireless scenario we illustrate possible effects on simulation results. As the ns-2 community relies heavily on exchanging hints and scripts, which keep re-infecting the knowledge-base even years after the introduction of the new RNG we believe that this might affect the majority of all ns-2 simulation results currently published.

1 Introduction

The ns-2 network simulator [1] is currently one of the most widely used environments for packet network simulation in the academic world. It is freely available in source code (in C++). Simulation scripts are written in OTcl, an object-oriented variant of the Tool Command Language (Tcl) [2].

Up until and including version 2.1b8 ns-2 used an implementation of the minimal standard multiplicative linear congruential generator by Park and Miller [3] for random number generation. This random number generator (RNG) has been shown to exhibit several weaknesses: first, it only offers a quite short-period length of only $p = 2^{31} - 2$, which can be a problem for long-running simulations, and second, it

M. Umlauft (✉)
Women's Postgraduate College for Internet Technologies, Vienna University of Technology, Favoritenstr. 9-11, A-1040 Vienna, Austria
e-mail: umlauft@big.tuwien.ac.at

S. Powell, J.P. Shim (eds.), *Wireless Technology*, Lecture Notes in Electrical Engineering 44, DOI 10.1007/978-0-387-71787-6_14,

is sensitive to the chosen seed as shown by Entacher and Hechenleitner [4]. When "bad" seeds are chosen, the created random variables will be correlated.

Therefore, in ns-2 version 2.1b9, a new RNG – the combined multiple recursive generator MRG32k3a proposed by L'Ecuyer [3–5] – was introduced to remedy these problems together with a new API (application programmer's interface). It is still used in all versions of ns-2 up to and including the current version 2.31.

Typically, in a simulation, a seed is set once in the simulation script depending on the number of the simulation run (the so-called "replication"). Afterwards the final result of the simulation is calculated by averaging over the trace output of several (hopefully many) replications. If the random numbers created by the RNG using these different seeds are correlated, this results in correlation between the output of those separate simulation runs which is, of course, undesirable.

The old minimal standard generator required to set seeds for random variables explicitly in the simulation script as shown in a typical example below.

```
set rng [new RNG]
$rng seed <n>
set e [new RandomVariable/Exponential]
$e use-rng rng
```

A new RNG object is created on line 1 and seeded in line 2 where `<n>` can be any positive integer or *0*. Setting the seed to *0* will seed the RNG with a different seed calculated from the computer's clock every time the simulation is started. Since the seed is not known in this case, most people use a positive integer value corresponding to the current replication number. In lines 3 and 4 an exponential random variable which uses this RNG object is set up. In the following, we call this method of setting the seed explicitly the "old method" or "old API." When the current MRG32k3a implementation is used this old method of setting the seed wrongly overrides the automatic seed generation of the new generator without giving any error message. Instead, *only the new API should ever be used with the new RNG implementation* as shown below (where `$rep` denotes the replication number).

```
set rng [new RNG]
for {set i 1} {$i < $rep} {incr i} {
   $rng next-substream;
}
set e [new RandomVariable/Exponential]
$e use-rng $rng
```

Unfortunately, the official ns-2 manual [6] can be misunderstood on the issue of correctly seeding the RNG. While it states "You should only set the seed of the default RNG." on p. 218 it still shows the old API functions for seed setting on p. 220, 223, and 226 without any warning that this compromises the seed insensitivity of the new MRG32k3a RNG. Also the popular and otherwise excellent lecture notes by Eitan Altman and Tania Jimenez [7] include several examples of old API function usage.

Since the ns-2 simulator is a complex program which has a steep learning curve, the ns-2 community relies heavily on exchanging tips and scripts, mostly

Table. 1 ns-users mailing list postings for "rng seed" from 2005–2007

Type of posting	#
Advice or example incorrectly using old method	29
Correct advice in response to seeding question	3
Example containing correct method in other context	7
Ambigous example or advice	4
Advice to use consecutively numbered seeds	2

number of postings.

on the ns-users mailing list [8]. Searching for "rng seed" on the mailing list archive search page http://www.isi.edu/cgi-bin/nsnam/htsearch gives 82 matches for the years 2005–2007 (up to and including July 2007) distributed as shown in Table 1.

Especially long-time users who re-use old simulation scripts containing the old seed setting method may fail to realize that *even though they are using the new implementation of the random number generator, they can still get correlated results*. In addition, by giving well-meaning advice on the mailing list sharing their old scripts, they propagate wrongful use of the old method to new users even many years after the introduction of the new RNG (the "allinone" distribution ns-2 version 2.1b9 is dated April 22, 2002).

Due to this lack of respective error messages, vague documentation and re-propagation of the old method on the ns-users mailing list, we believe that this might affect up to 80% of all ns-2 simulation results currently published.

The chapter is organized as follows: in Section 2 we give an overview of the theoretical background of the MRG32k3a random number generator and show the general effects of wrongly using the old method to seed the RNG. Practical effects are illustrated in Section 3 by means of two example scenarios – a simple wired topology (Section 3.1) and a wireless multi-hop network (Section 3.2). In Section 4 we discuss the root of the problem by inspecting the source code of ns-2 and give advice on how to avoid it. Finally, Section 5 concludes the chapter with some thoughts on the impact on currently published simulation results.

2 The ns-2 Random Number Generator

As stressed in works such as [9] and [10] the reliable and efficient provision of random numbers is a core prerequisite for any network simulator tool. In the following we give a short overview of the theoretical background of the ns-2 MRG32k3a RNG and show the general effect of using the old API with the new RNG by investigating correlation between unform random variables created with ns-2.

2.1 Theoretical Background

Mathematically speaking, a random number generator (RNG) consists of a deterministic algorithm which aims at producing a sequence of numbers that is statistically indistinguishable from real random events as produced, e.g., by means of radioactive decomposition. More precisely, we deal with the realization of i.i.d. U(0,1) random variables u_0, u_1, u_2...for which the *t*-dimensional vector

$$u_{n,t} = (u_n, u_{n+1}, \ldots, u_{n+t-1}) \tag{1}$$

is uniformly distributed over the *t* - dimensional hypercube $(0,1)^t$ for all integers $n \geq 0$ and $t > 0$ [11].

The classical approach to define a RNG makes use of a linear recurrence of the general form

$$x_n = (a_1 x_{n-1} + a_2 x_{n-2} + \ldots + a_k x_{n-k}) \bmod m \tag{2}$$

where the integer *k* is referred to as the order of the RNG and the integer *m* is called the modulus. The resulting sequence of numbers $u_n = x_n/m$ is periodic, however, a clever choice of the integer coefficients $a_1, \ldots a_k$ allows to achieve periods of maximal length which is of course one of the main quality criteria of such a so-called multiple recursive generator (MRG) [11].

As a simple example, choosing $k = 1$ yields the class of linear congruential generators (LGC) which have been widely used in simulation tools until a couple of years ago, usually with $m = 2^{31} - 1$. However, the period of an LGC is limited by $m - 1$ which is far too short to be sufficient for current network simulation scenarios. Therefore more recently it has been proposed to combine several MRGs towards so-called "combined multiple recursive generators." Putting it simple, this new RNG class consists of normalized linear combinations of *J* copies of ordinary multiple recursive generators $x_{j,n}$ of order *k* ($j = 1, \ldots, J$)

$$x_{j,n} = \left(a_{j,1}x_{n-1} + a_{j,2}x_{j,n-2} + \ldots + a_{j,k}x_{j,n-k}\right) \bmod m_j \tag{3}$$

with distinct primes m_j and $a_{j,l}$ being integers between 0 and m_j. It has been argued [11] that these combined MRGs provide a very efficient way of implementing an MRG with large modulus *m* and several large coefficients.

The most prominent example of a combined MRG is the MRG32k3a random number generator due to L'Ecuyer [5]. MRG32k3a has $J = 2$ components of order $k = 3$ with carefully chosen coefficients and moduli as shown in Eqs. (4) and (5).

$$x_{1,n} = \left(1403580x_{1,n-2} - 810728x_{1,n-3}\right) \bmod \left(2^{32} - 209\right) \tag{4}$$

$$x_{2,n} = \left(527612x_{2,n-1} - 1370589x_{2,n-3}\right) \bmod \left(2^{32} - 22853\right) \tag{5}$$

Combining $x_{1,n}$ and $x_{2,n}$, the eventual output u_n is then defined as

$$u_n = \frac{\left(\left(x_{1,n} - x_{2,n}\right) \bmod \left(2^{32} - 209\right)\right)}{2^{32} - 208} \tag{6}$$

MRG32k3a has a period length of approx. 3.1×10^{57} and has been demonstrated to behave well for a broad range of statistical test scenarios. In [12], it is shown how this generator can be further generalized for producing multiple streams and substreams of random numbers. To this end, it is proposed to cut the resulting (long) sequence of random numbers into adjacent streams of length $Z = 2^z$ and then partition each such stream into 2^v substreams (blocks) of length $W = 2^{z-v}$. Note that according to [6], the choice $Z = 2^{127}$ is particularly suitable as the corresponding spectral tests demonstrate the high level of statistical independence between the corresponding (sub-)streams. The resulting generator provides 1.8×10^{19} independent streams of random numbers, each of which consists of 2.3×10^{15} substreams with a period of 7.6×10^{22} each. In ns-2, each of these substreams corresponds to an individual RNG object, hence on creation of a new RNG object, simply the next substream is used.

In order to start the MRG32k3a, we need initial values for each of the six variables $\{x_{1,0}, x_{1,1}, x_{1,2}; x_{2,0}, x_{2,1}, x_{2,2}\}$ which can conveniently be described as a six-dimensional "seed vector." It is crucial to note that the nearly perfect randomness of the entire (long) sequence is of course maintained approximately also on a stream and substream level and thus for every RNG object, whereas setting explicitly a new seed vector for a newly created RNG object only could destroy this extremely desirable insensitivity property.

2.2 General Effects

To test the effect of using the old API with the new RNG we implemented a simple correlation experiment. For simplicity of visualization we chose to conduct our experiment with three uniformly distributed random variables. If we interpret the values drawn from these variables as a vector in 3D space, we expect to see a uniform "cloud" of points if the variables are uncorrelated as shown in Fig. 1.

We use the following simulation script to set up three uniform random variables using the old API functions:

```
for {set i 0} {$i < 3} {
    set rng($i) [new RNG]
    $rng($i) seed $n($i)
    set u($i) [new RandomVariable/Uniform]
    $u($i) use-rng $rng($i)
}
```

We use one RNG object with its own seed for each random variable using different sets of seeds `$n($i)` as shown in Table 2. We then interpret the values drawn for the random variables `$u(1)`, `$u(2)`, and `$u(3)` as a vector and plot them for

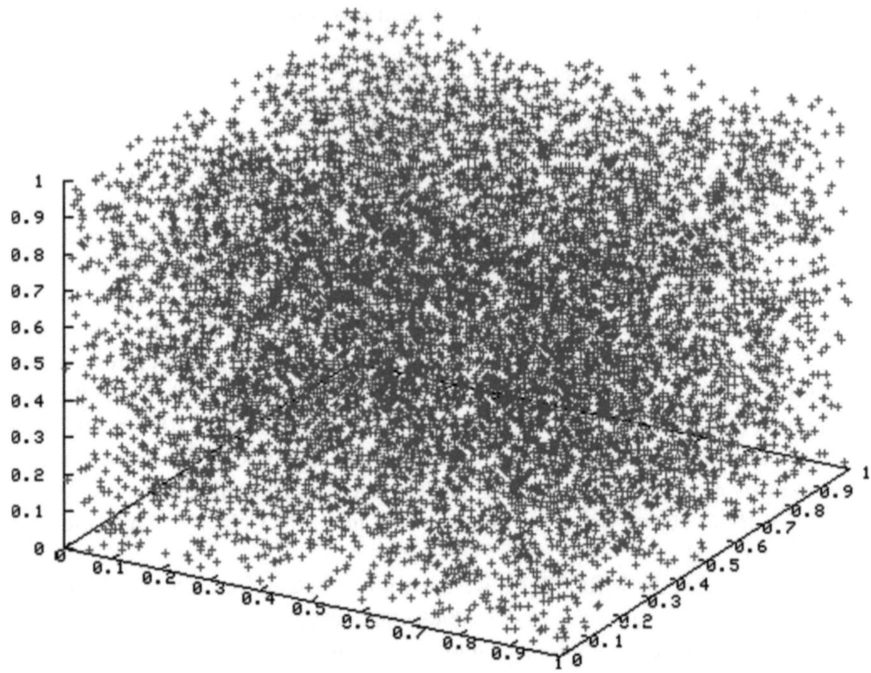

Fig. 1 Values from three uncorrelated uniformly distributed variables interpreted as vector in 3D space, 10,000 values drawn

Table 2 Sets of seeds

Random variable/seed	Set 1 ("good")	Set 2 ("bad")	Set 3 ("bad")
`$u(1)/$n(1)`	1973272912	1	1
`$u(2)/$n(2)`	1822174485	2	634005912
`$u(3)/$n(3)`	1998078925	3	634005911

the new MRG32k3a RNG and the old Park/Miller RNG. We also plot the results for the new RNG using the new seed setting method and for the old RNG using a set of known "good" seeds. The seed values are taken from [4] with set 1 being a set of known "good" seeds and sets 2 and 3 consisting of known "bad" seeds for the old Park/Miller RNG.

Figure 2 compares the random values generated by the old Park/Miller RNG using the known "good" seed set 1 versus the current MRG32k3a RNG using the new seeding method (new API).

Results for known "bad" seed sets 2 and 3 are shown in Figs. 3 and 4, respectively. While the actual numbers generated are different for the MRG32k3a RNG and the Park/Miller RNG, we can see that the behavior is similarly bad for "bad" seed choices (the difference is not noticeable at the resolution of the figures and due to the differing periods of the two generators).

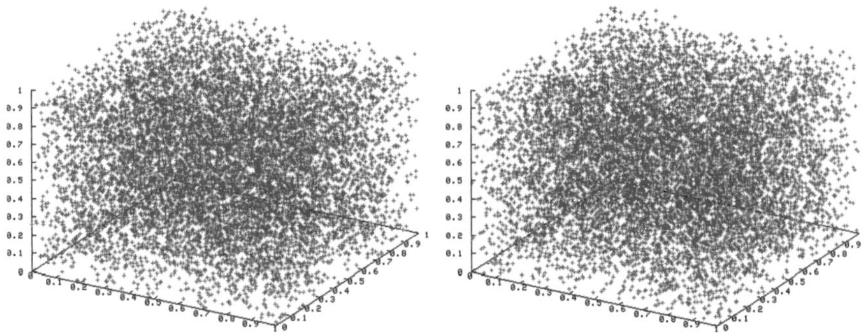

Fig. 2 No correlation, 10,000 values drawn, *left*: old Park/Miller RNG with known "good" seed set 1, *right*: MRG32k3a using new API (© 2007 IEEE)

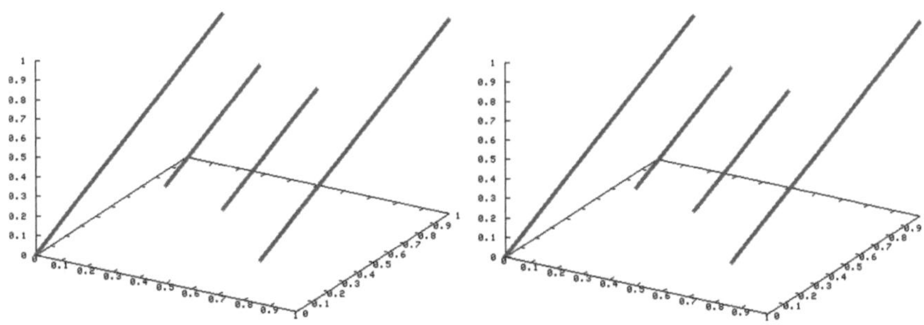

Fig. 3 Correlation with known "bad" seed set 2, 10,000 values drawn, *left*: old Park/Miller RNG, *right*: MRG32k3a (© 2007 IEEE)

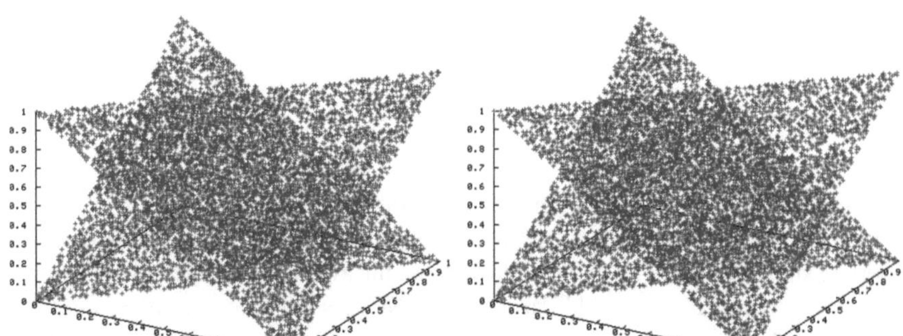

Fig. 4 Correlation with known "bad" seed set 3, 10,000 values drawn, *left*: old Park/Miller RNG, *right*: MRG32k3a (© 2007 IEEE)

3 Practical Effects on Network Simulation

To illustrate the effect on network simulation results we investigate two simple scenarios, a wired topology and a wireless example. While these examples are chosen for simplicity and use very simple models, effects can still be shown.

3.1 Scenario 1: Wired Topology

Inspired by [4] we generate one of the simplest wired topologies, two nodes linked by a bottleneck link (see Fig. 5). Node 1 has a DropTail Queue with a maximum size of 1,000 packets. G1 to G5 are exponential traffic generators (class `Application/Traffic/Exponential`) generating on/off traffic.

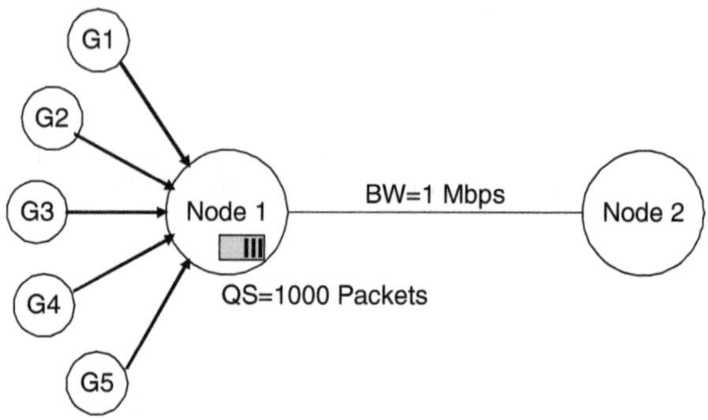

Fig. 5 Wired topology – bottleneck link (© 2007 IEEE)

During each on-interval one packet with a size of 1000 bytes and an on-time of 0.08 μs is generated (resulting in an internal "rate" of 1 Gbps). The mean of the exponentially distributed off-time is set to 41 ms. Therefore, the average arrival rate is $\lambda = \mathit{8000\ bits/41\ ms} = \mathit{0.195\ Mbps}$ for each generator and $\Sigma\lambda = \mathit{0.976\ Mbps}$ total. This gives a utilization factor of $\rho = \Sigma\lambda/BW{=}0.976$. Calculating the mean queue length as

$$\overline{q} = \frac{\rho}{1-\rho} - \frac{\rho^2}{2(1-\rho)} \tag{7}$$

we expect an average queue length of $\overline{q} = 20.488$packets.

Simulation time was 7,200 s with a sampling interval of 10 ms. The used sets of seeds are shown in Table 3, while Table 4 shows the average queue lengths measured for the new MRG32k3a and old Park/Miller RNG. As we have seen in Section 2.2, when wrongly using the old API with a "good" set of seeds still yields uncorrelated results. Therefore, since an error will be incurred by unsuspecting users if the new RNG is seeded using the old API *and* a "bad" set of seeds we compare simulation results for the new API vs. the old API with a known "bad" seed set for the new RNG. We also show results for a known "good" seed set vs. a known "bad" seed set for the old RNG in comparison. Figure 6 shows the empirical CDF of the queue lengths. As can be seen, the use of the "bad" seed set leads to higher values for the average queue length and generally longer queues while with the new API or a known "good" seed set in case of the old RNG the values are close to each other and to the theoretical result.

Table 3 Sets of seeds

Generator	Set 1 ("good")	Set 2 ("bad")
G1	1973272912	1
G2	1822174485	2
G3	1998078925	3
G4	678622600	4
G5	999157082	5

Table 4 Average queue lengths

RNG	Seed set	Average queue length
new MRG32k3a	New API	20.2996
new MRG32k3a	2 ("bad")	29.4527
old Park/Miller	1 ("good")	19.4398
old Park/Miller	2 ("bad")	24.2785

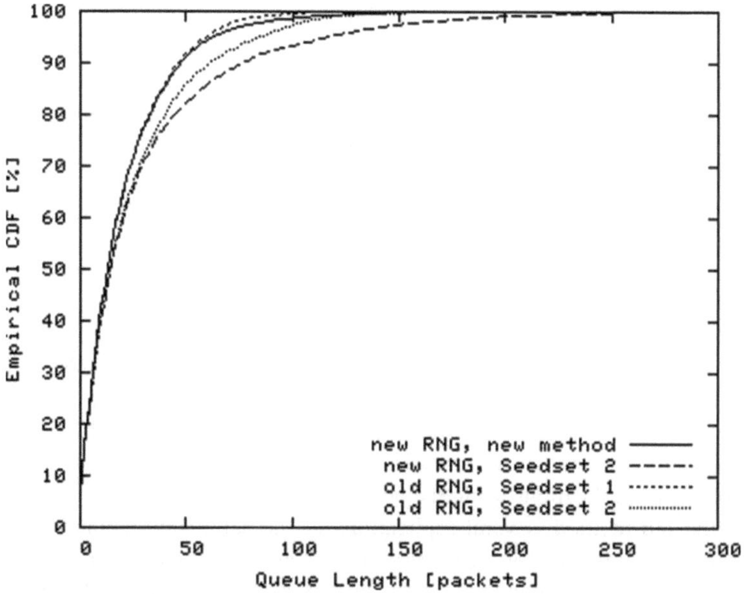

Fig. 6 Empirical CDF of queue lengths (© 2007 IEEE)

3.2 Scenario 2: Wireless Multi-Hop Topology

Scenario 2 is a wireless multi-hop network consisting of four nodes connected by wireless links with a capacity of 1 Mbps configured in a line topology (see Fig. 7). For reasons of simplicity, we model the wireless channel with a uniform error model by installing `ErrorModel` objects as `lossmodel` on the links between the nodes.

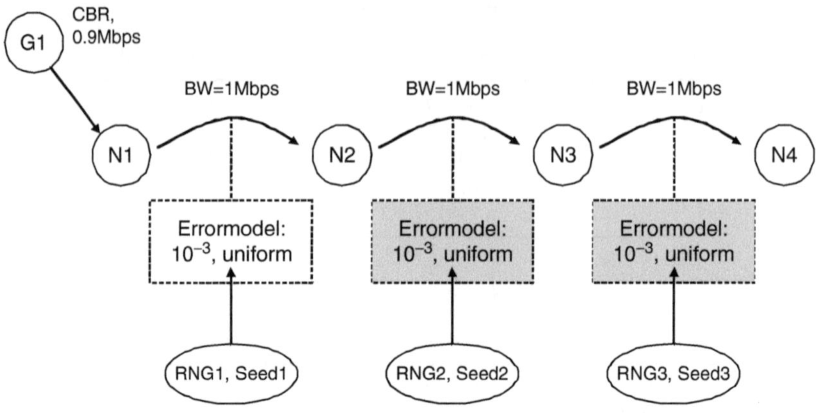

Fig. 7 Wireless multi-hop network, line topology (© 2007 IEEE)

Packets on the links are randomly dropped with an error rate of 10^{-3}. Each link has its own errormodel with its own RNG. While we are aware that a uniform error model does not reflect the reality of a wireless channel well (as errors are bursty for wireless media) this does not matter as the aim of the example is just to show that the experienced burstyness of errors for a multi-hop transmission is changed when a bad method of seeding the RNG is used.

We configure a single constant bit rate (CBR) traffic source with a rate of 0.9 Mbps on node N1 which sends traffic from node N1 via intermediate nodes N2 and N3 to the sink N4 using the UDP protocol. Since the links have a capacity greater than the traffic rate all packet loss occurs due to errors on the links.

Simulation time is 600 s (corresponding to a 10 min flow) for each replication.

We investigate the effect of wrongly using the new MRG32k3a RNG with the old API and the known "bad" seed set as opposed to correctly using the new API. The known "bad" seed set is constructed from seed set 2 shown in Table 3 using the first three seeds (1, 2, and 3) for RNGs 1, 2, and 3, respectively. This is compared to the results of 10 replications using the new method.

With the old method, several replications of the same simulation were differentiated by explicitly setting different seeds for every replication. In contrast, the API for the new method offers the `next-substream` function to set up (seed) the RNG according to the replication number. To seed an RNG for, say, the fifth replication one simply has to call the `next-substream` function of the RNG object five times. An example is given in the code below which sets up three RNGs according to the current replication (given in `$rep`).

```
for {set i 1} {$i < $rep} {incr i} {
   $rng1 next-substream;
   $rng2 next-substream;
   $rng3 next-substream;
}
```

For each transmission over the multi-hop network we observe bursts of errors and the so-called "runs" of good packet transmissions. Figure 8 shows the empirical CDF of the lengths of good packet runs. As can be seen, all 10 replications generated with the new method yield quite similar results. In contrast the result generated with the known "bad" seed set 2 differs significantly, showing slightly more short and many more medium sized (24–220 packet long) runs. For short runs of 6–23 error-free received packets the difference is only between 1 and 4%, but for run-lengths between 24 and 220 packets the difference is more pronounced with up to almost 19% more 120 packet long runs. On the other hand, the number of really long runs of 1,000 packets or more is slightly lower than when the correct new method of setting up the RNG is used. This means that when the old API is used in combination with a "bad" seed set, results are skewed in favor of medium sized runs of error-free received packets.

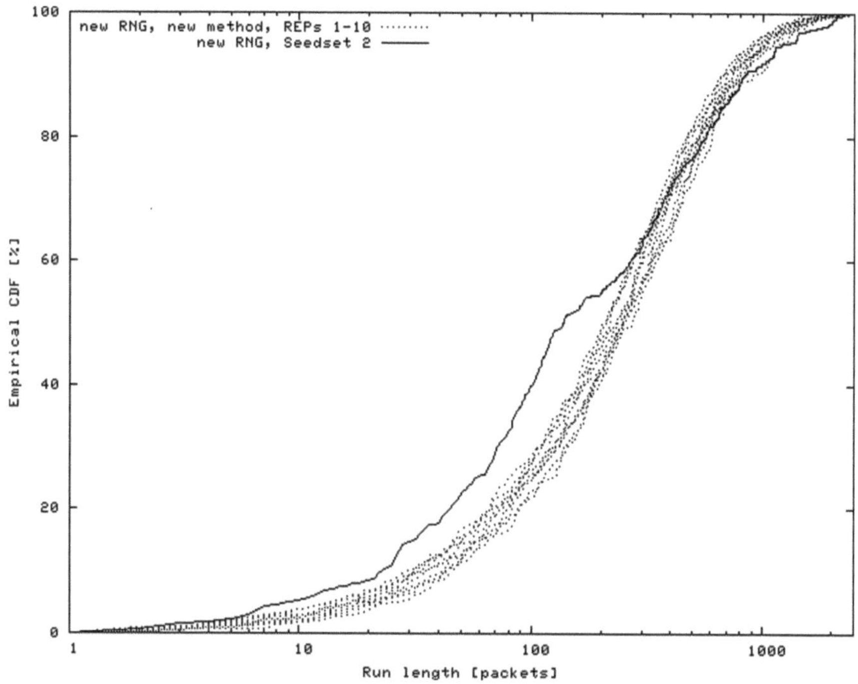

Fig. 8 Empirical CDF of run-lengths of error-free-received packets (© 2007 IEEE)

4 Getting it Right

To gain an understanding of why the problem occurs in the first place, we inspect the source code of ns-2, version 2.31 from the "allinone" distribution. The random number generator is implemented in the files `rng.cc` and `rng.h` in the `ns/tools` directory. These files contain both implementations, one for the current MRG32k3a RNG and one for the old Park/Miller RNG. The user can choose to enable the old Park/Miller RNG by setting the define directive `#define OLD_RNG` in `rng.h` and recompiling. By default, the MRG32k3a RNG is enabled.

When a simulation script written in OTcl is executed, explicitly setting the seed with the command `$rng seed <n>` (`<n>` being a non-negative integer greater than zero) is processed in the C++ code in `rng.cc` on line 236 by the `RNG::command` () method (which starts on line 219). This calls `set_seed (RAW _SEED _SOURCE, s)` with `s` being the seed `<n>` passed from the OTcl script. The method `RNG::set _seed` () is implemented starting on line 325. If the MRG32k3a RNG is used, this will call `set _seed (seed)` on line 378. This, in turn implemented starting on line 808 where it creates a six-element vector where all elements are set to the seed value (iow. all six elements of the vector are now of the same value equal to the seed explicitly set in the simulation script). Eventually (line 818) it calls `RNG::set _package _seed` () which passes the elements of this vector into the six members of `next _seed _`, the package-wide six-dimensional seed vector. It is a static member of the `RNG` class; iow. it exists only once for *all* objects of type `RNG` and is shared among them. On creation, each new `RNG` object will use the seed vector to seed itself and then recalculate `next _seed _` (line 754 ff.) to set it up for the next `RNG` object to be created (compare calculation of $x_{j,n}$ in Section 2.1).

Therefore, seeding each new `RNG` object explicitly in the simulation script overrides the seed vector calculation mechanism of MRG32k3a and hard-seeds the RNG to the values given in the OTcl script! As demonstrated in Section 2.2, this leads to correlation among random variables when bad seeds are chosen.

This problem can be avoided by *only ever using* the new method to seed the RNG using the `next-substream` API function instead of seeding each of the RNG objects explicitly.

Example code is shown below and can also be found on pp. 217 f., Section 24.1.1 in [6].

```
set rng1 [new RNG]
set rng2 [new RNG]
for {set i 1} {$i < $rep} {incr i} {
   $rng1 next-substream;
   $rng2 next-substream;
}
set u1 [new RandomVariable/Uniform]
set u2 [new RandomVariable/Uniform]
$u1 use-rng $rng1
$u2 use-rng $rng2
```

The above sets up two RNG objects according to the current replication given in `$rep`. Then, two uniform random variables are attached to these RNGs. Random numbers drawn from these variables will not be correlated.

Optionally, the `defaultRNG` object (but *none* of the other RNG objects) may be seeded using the code shown below.

```
global defaultRNG;
$defaultRNG seed <value>
```

5 Conclusion

The reliable and efficient provision of random numbers is a core prerequisite for any network simulator tool. To improve on the quality of generated random numbers, espically to increase the period and to remedy the problem of sensitivity to seeds the old minimal standard multiplicative linear congruential generator by Park/Miller has been replaced with the combined multiple recursive generator MRG32k3a by L'Ecuyer in the ns-2 network simulator as of version 2.1b9 and is still used in the current version 2.31.

In the course of this change, also the API has been changed from simply setting the seed for the RNG explicitly to invoking a function call several times. Inspecting the source code of ns-2 we find that using the old method to explicitly set seeds for the current MRG32k3a results in overwriting of the package-wide seed of the generator, thereby confounding the new, automatic seed-generation mechanism implemented even though no error message is thrown. If bad seeds are chosen, this leads to correlation between the generated random variables. Iow. Even when using the new RNG with the old API one can experience similarly bad behavior as with the old RNG!

Unfortunately, not only can the ns-2 manual be misunderstood on how to correctly seed the new RNG, also other popular tutorials and the majority of postings on the ns-users mailing list give outdated and therefore harmful advice. The occurrence of outdated advice on the ns-users mailing list is especially harmful because the ns-2 community relies heavily on the exchange of hints and scripts to remedy the steep learning curve of the simulator. In particular, experienced users who well-meaningly share their old simulation scripts on the list re-infect the knowledge base over and over again. For the years 2005–2007 (up to and including July) 35 incorrect vs. only 10 correct postings on the matter were found even though the new RNG was already introduced in April 2002.

Therefore, we believe that a very large number (maybe even the vast majority) of ns-2 simulation results currently published is based on scripts using an incorrect method to seed the RNG. Out of those, the number of actually affected results is hard to estimate as it depends on several other factors: the choice of seed and how the RNG objects are used. Note that when "good" seeds are chosen the RNG still behaves well even though the old API is used. Results most prone to correlation are those which use several RNG objects seeded with different seeds within a single simulation run, especially when consecutively numbered seeds are used.

The problem can be completely avoided by *using only the new API* to seed the RNG which we strongly recommend!

Acknowledgments This research has been partly funded by the Austrian Federal Ministry for Education, Science, and Culture, and the European Social Fund (ESF) under grant 31.963/46-VII/9/2002 and partly by the Austrian K*plus* competence center program. Figures 2–8 reprinted from [13] with kind permission by IEEE.

References

1. *nsnam web pages*, http://www.isi.edu/nsnam/, last visited: Jul. 2007.
2. K. John: Ousterhout: *Tcl and the Tk Toolkit*, Addison-Wesley, Reading, MA, USA, ISBN 0-201-63337-X, 1994.
3. S.K. Park and R.W. Miller: Random number generation: Good ones are hard to find. *Communications of the ACM*, 31(10), 1192–1201, Oct. 1988.
4. B. Hechenleitner and K. Entacher: On Shortcomings of the ns-2 Random Number Generator. In T. Znati and B. McDonald, eds., *Communication Networks and Distributed Systems Modeling and Simulation (CNDS)*, 2002.
5. P. L'Ecuyer: Good parameters and implementations for combined multiple recursive random number generators. *Operations Research*, 47(1), 159–164, 1999.
6. K. Fall and K. Varadhan (Eds.): *The ns Manual (formerly ns Notes and Documentation)*, http://www.isi.edu/nsnam/ns/ns-documentation.html, last visited: Oct. 2006.
7. E. Altman and T. Jimenez: *ns-2 for Beginners*, lecture notes, Dec. 2003, http://www-sop.inria.fr/maestro/personnel/Eitan.Altman/COURS-NS/n3.pdf, last visited: Oct. 2006.
8. ns-users mailing list, ns-users@isi.edu, subscription on the Web via http://mailman.isi.edu/mailman/listinfo/ns-users, last visited: Jul. 2007.
9. K. Pawlikowski, H.-D.J. Jeong, and J.-S.R. Lee: On Credibility of Simulation Studies of Telecommunication Networks. *IEEE Communications Magazine*, 40(1), 132–139, Jan. 2002.
10. S. Kurkowski, T. Camp, M. Colagrosso: *MANET Simulation Studies: The Incredibles*, ACM SIGMOBILE Mobile Computing and Communications Review, 9(4), Oct. 2005.
11. P. L'Ecuyer: Random Number Generation, *Chapter 2 of the Handbook of Computational Statistics*, J.E. Gentle, W. Haerdle, and Y. Mori, eds., Springer-Verlag, New York, 2004, 35–70.
12. Pierre L'Ecuyer et al.: An object-oriented random number package with many long streams and substreams. *Operations Research*, 50(6), 1073–1075, 2002.
13. M. Umlauft and P. Reichl: Experiences with the ns-2 network simulator – explicitly setting seeds considered harmful. In *Proc. Wireless Telecommunications Symposium (WTS)*, April 2007, Pomona, CA, USA.

Topology-Based Routing for Xmesh in Wireless Sensor Networks

Lei Wang and K. Wendy Tang

Abstract Large wireless sensor networks can contain hundreds or thousands of sensor nodes. Due to wireless sensor network's properties of low-energy-efficiency, large-scale, low cost, and lossy nature, the development of efficient routing protocols for these large and dense wireless sensor networks is an interesting research topic. This research focuses on the design and implementation of protocols for dense and wireless sensor networks. More specifically, we propose to combine an underlying topology with *Xmesh,* the multihop routing strategy of Crossbow Technology's motes. In an effort to limit the path lengths, we propose to impose an underlying connectivity graph for *Xmesh*. The underlying connectivity graph is a virtual topology of the network, hence the name "topology-based routing." Instead of being always forwarded to the node with the best link quality among all neighbors, a packet is being routed according to the shortest path routing of the underlying graph which guarantees a bounded path length. *Cayley* graphs from the *Borel* subgroup have been known as the densest degree-4 graphs and all *Cayley* graphs are vertex transitive. In this work, we propose a topology-based routing for *Xmesh* with *Cayley* graphs as the underlying topology. We show that, indeed, by imposing a *Cayley* graph as an underlying graph, the average path lengths between nodes is smaller and that the averaged power consumed is less than the original *Xmesh*.

1 Introduction

Wireless sensor network (WSN) consists of a large number of nodes with different kinds of sensors, linked by a wireless medium (radio frequency) to perform distributed sensing tasks. These networks can be applied in many environments such as intelligent battlefields, smart hospitals, environment response systems, and

---

L. Wang (✉)
State University of New York, Stony Brook, New York, USA
e-mail: leiwang@mail.ee.sunysb.edu

S. Powell, J.P. Shim (eds.), *Wireless Technology*, Lecture Notes in Electrical Engineering 44, DOI 10.1007/978-0-387-71787-6_15,

surveillance systems. In most applications, mainly unwired power supply and communication bandwidth are constrained for sensor nodes [1]. Therefore, to shorten network lifetime and to use the limited bandwidth efficiently, researchers emphasize energy conservation in the design and management of WSNs [2]. At the network layer, finding methods for energy-efficient route discovery and relaying of data from the sensor nodes to the base station is highly desirable. There are still other concerns when designing WSN protocols, such as fairness, fault tolerance, node/link heterogeneity, and network dynamics. The dynamic and lossy nature of wireless communication poses major challenges to reliable, self-organizing multihop networks. For large and dense WSN with a few hundred or thousands of nodes, energy conservation, scalability, and self-configuration are the primary goals [3, 4]. Many protocols have been proposed for WSN [1–6].

Crossbow Technology Inc. has been one of the major vendors for wireless sensor networks. Its powerful battery-powered platform runs on the open-source TinyOS operating system. With this operating system, developers can control low-level event and maintain task management. Its multihop routing protocol called "Xmesh" is a distributed routing process [7, 8]. Routing decisions are based on a minimum transmission cost function that considers link quality of nodes within a communication range. However, there are no limits on the path length. In extreme cases and for large networks, it is conceivable that a packet may need to hop through many intermediate nodes before reaching its intended destination.

In an effort to limit the path lengths, we propose to impose an underlying connectivity graph for Xmesh. The underlying connectivity graph is a virtual topology of the network. Instead of being forwarded to the best link quality node among all neighbors within communication range, a packet is being routed according to the shortest path routing of the underlying graph. In the event that multiple shortest paths exist, the one with the best link quality is chosen. The purpose of the underlying connectivity graph is to impose a virtual topology that facilitates routing and guarantees a bounded path length. An ideal underlying graph should guarantee a small number of hops between nodes and should possess a simple routing algorithm.

Cayley graphs from the Borel subgroup [9] have been known as the densest degree-4 graphs and all Cayley graphs are vertex transitive or symmetric. Furthermore, our earlier work has shown that Cayley graphs can have very effective integer representation and symmetric routing strategies [9–12]. In this chapter, we propose a topology-based routing for Xmesh with Cayley graphs as the underlying virtual topology. To evaluate the effect of imposing such a virtual topology on Xmesh, we simulated our proposed protocol via Power Tossim, an emulator for wireless sensor networks. We show that, indeed, by imposing a Cayley graph as an underlying graph, the average path lengths between nodes is smaller and that the averaged power consumed is less than the original Xmesh.

This chapter is organized as follow: Section 2 provides an overview of Cayley graphs and its routing algorithm. Section 3 gives a detailed description of our proposed topology-based routing for Xmesh. Simulation results and analysis are described in Section 4. A summary and conclusions are described in Section 5.

2 Routing for Cayley Graphs

2.1 Cayley Graph Overview

Symmetric, regular, undirected graphs are useful models for interconnection of multicomputer systems. Dense graphs of this sort are particularly attractive. Based on group theoretic constructions, Cayley Graphs [13] are in this category of graphs [9]. The construction of Cayley graphs is described by finite (algebraic) group theory.

Definition *A graph* **C** = (**V**, **G**) *is a Cayley graph with vertex set* **V** *if two vertices* v1; v2∈**V** *are adjacent* ⇔v1 = v2∗g *for some* g∈*G where* (**V**,∗) *is a finite group and* **G**⊂**V**\{I}. **G** *is called the generator set of the graph.*

Note that the identity element *I* is excluded from G. This prevents the graph from having self-loops. In this chapter, we are interested in undirected, degree-4 Cayley graphs. In other words, the generator set consists of two group elements and their inverses. Cayley graphs are vertex transitive. Furthermore, the densest known degree-4 graphs are Cayley graphs from the Borel group [13].

The dense property of Cayley graphs implies that they can connect a large number of nodes via a small number of hops through intermediate nodes. The vertex-transitive property is useful for routing. It means that a Cayley graph "looks the same from any node" which means the same routing algorithm and routing table can be used at every node. More specifically, routing between vertices *i* and *j* can be determined by finding paths between vertices *0* and *j'*, where *j'* is a function of *i* and *j*. This property is the basis for a distributed routing algorithm, the vertex-transitive routing in [10–12]. Figure 1 is a 21–node, degree-4, Borel Cayley graph in the integer domain.

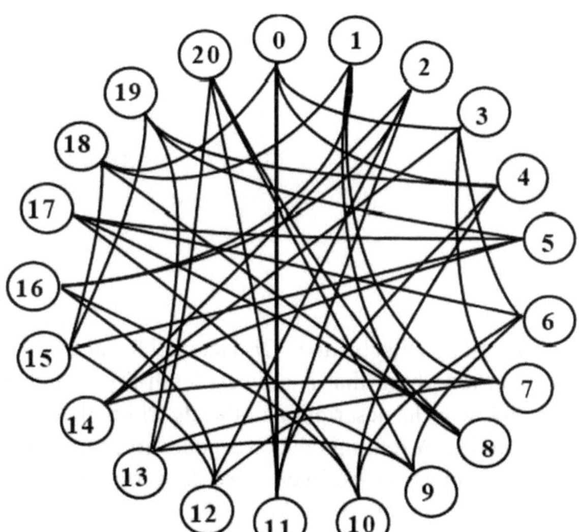

Fig. 1 A 21-node Cayley graph

The graph has V = {0, 1, . . ., 20}, and the connection is defined as [9–12]:

Let V = {0, 1, . . ., 20}. For any i∈V , if i mod 3 = :
0: i is connected to i+3, i−3, i+4, i−10; mod 21
1: i is connected to i+6, i−6, i+7, i−4; mod 21
2: i is connected to i+9, i−9, i+10, i−7; mod 21

2.2 Shortest Path Algorithm for Cayley Graph

To implement Cayley Graph into Xbow's motes, we implemented an algorithm for path searching to find the shortest path between any two vertices. In [5], each vertex has 4 degree, 2 for incoming communication and 2 for outgoing communication. In this chapter, the two incoming and two outgoing channels merge into one channel for single half duplex transceiver. Therefore, there are multiple choices of shortest path selecting in terms of hops. For example, in Fig. 2, there are two shortest paths (with a hop count of three) from noted 0 to node 9.

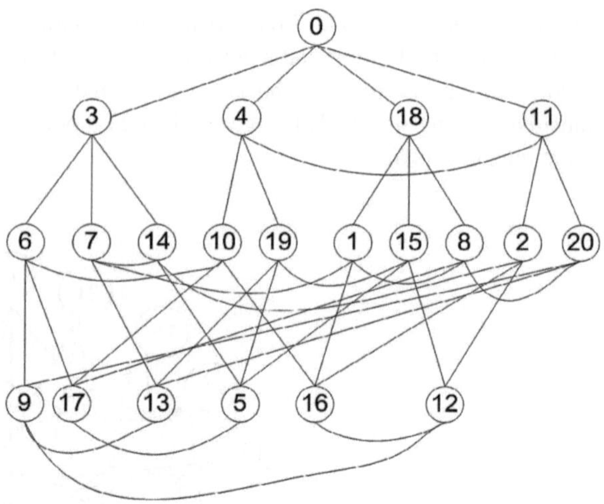

Fig. 2 Tree view with node 0 as root

Inspired by Dijkstra and "On-Demand Route Discovery" algorithms [14, 15], we chose an algorithm that combines depth-first-search [16] and breadth-first-search [17]. Below is the description of the algorithm that we use to generate routing tables for all nodes. A routing table at a node stores the optimal outgoing links from that node to all other nodes. Because of the vertex-transitive property of Cayley graphs, routing tables for all the nodes can be generated by a simple integer mapping of the routing table from node 0 Table 1 [9–12].

Table 1 An algorithm to generate routing tables for all nodes

- Step 1: Initialize the four possible paths and corresponding distance of each node with infinite numbers.
- Step 2: Starting from the four nodes j directly connected to node 0 (distance = 1), Breadth-first search the nodes below these four nodes with the recursive function.
  1) While the depth of searching ≤ log4(n)+ 1, do 2).
  2) If next connected node j1/j2/j3/j4 is not visited through this link a/b/c/d, in other words, distance = infinity, and j1/j2/j3/j4 is not root of this tree, go to the following procedure:
     a) Update the distance and path information for node j1/j2/j3/j4;
     b) Find the next four connected nodes of node j1/j2/j3/j4, flag this link a/b/c/ /d, call procedure 1).
- Step 3: Among the paths of each node, select the those with minimum distance, and flag those corresponding links in routing table for the tree with node 0 as root.
- Step 4: Manipulating the vertex transitive formula given by [12, 22, 23], generate the routing table for other vertices.

3 Topology-Based Routing with Xmesh

3.1 Introduction to Xmesh

The multihop routing protocol, Xmesh [7], is implemented in a TinyOS application called Surge. Xmesh is a distributed routing process that has three local processes: *link quality estimation*, *neighborhood management*, and *connectivity-based route selections*. The link quality estimator estimates the link quality of all nodes within communication range. The neighborhood management process decides how the node chooses neighbors for paths. Link estimation and neighborhood management build a probabilistic connectivity graph. The routing process then builds topologies upon this graph. These three processes together form a holistic approach with the goal of minimizing total cost and providing reliable communications. The core component of Xmesh is the neighbor table which contains status and routing entries for neighbors; its fields include MAC address, routing cost, parent address, child flag, reception (inbound) link quality, send (outbound) link quality, and link estimator data structures.

The Component of Parent selection is running periodically to select one of the potential neighbors for routing. A packet has fields for parent address, estimated routing cost to the base, and a list of reception link estimations of neighbors. When a node receives a route message already in its neighbor table, the corresponding entry is updated. If not, the neighbor table manager decides whether to insert the node or drop the update. Originated data packets, such as outputs of local sensor processing, are queued for sending with the parent as the destination. Incoming data packets are selectively forwarded through the forwarding queue. To avoid cycles the corresponding neighbor table entry is flagged as a child in parent selection. Duplicate forwarding packets are eliminated. When cycles are detected on forwarding packets, parent selection is triggered with the current parent demoted to break the cycle.

3.2 Topology-Based Routing with Xmesh

In our proposed topology-based routing with Xmesh, we use a Cayley graph as an underlying topology for Xmesh routing. In Xmesh, a packet is forwarded to an intermediate node that has the best link quality within the communication range of the source node. The packet will be forwarded until it reaches its intended destination. In extreme cases and for large networks, such multihop routing can result in long path length, i.e., a packet will go through large number of intermediate nodes.

By routing packets in Xmesh according to an underlying graph, we can impose a limit on the path length. In our proposed topology-based routing, a packet is forwarded to an intermediate node that has the best link quality *among neighbors of an underlying graph.* By imposing a topology on forwarding packets, the path length of the message is bounded by the *diameter* (maximum of the minimal distance) of the graph. Obviously, it is important to choose an underlying graph that is *dense* (small diameter for large number of nodes). We choose to use Cayley graphs as the underlying topology because of its vertex-transitive and potentially dense properties.

The vertex-transitive property of Cayley graphs enables us to use the same routing table at each node. A routing table is first generated off-line that stores the *optimal* outgoing links from node 0 to all other nodes in the network. An outgoing link is *optimal* if it contributes to a shortest path between the source and the destination. This routing table of size $O(n\delta)$, where n is the number of nodes and δ is the degree of the graph, is stored at every node. A vertex-transitive mapping formula [9–12] is used to identify the appropriate entry in the routing table to determine what are the optimal outgoing links.

4 Simulation Results and Analysis

4.1 Power TOSSIM

We implement the topology-based routing (based on Cayley graphs) for Xmesh in the TinyOS sensor network emulator, Power TOSSIM [18]. It is an extension to TOSSIM, a scalable simulation environment for wireless sensor networks, that provides an accurate, estimation of node power consumption. Power TOSSIM can capture the detailed, low-level energy requirements of the CPU, radio, sensors, and other peripherals based on the Mica2/Mica2dot/MicaZ sensor node platform of Crossbow Technology Inc [19]. Our implementation is based on Power/Radio Model of Mica2 in Power TOSSIM and the operating frequency is 915 MHz

4.2 Simulation and Results

To evaluate the effect of topology-based routing for Xmesh, we compare the performance of three different routing strategies: *original Xmesh*, *Xmesh based on Cayley graphs and shortest path routing*, and *Xmesh based on Cayley graphs but with random selection of routes*. For the original Xmesh, a packet is forwarded to intermediate nodes with the best link quality. For Xmesh based on Cayley graphs, a Cayley graph is the virtual topology of nodes in the network, packets are forwarded to intermediate nodes that is part of the shortest path between the source and the destination. In the event that multiple shortest paths exist, the best quality link will be chosen. For Xmesh based on Cayley graphs and random selection of routes, a Cayley graph is still the virtual topology of the networks, but packets are randomly forwarded to one of the neighbors of the virtual topology.

As an example, for a 21-node network, a packet is being sent from source node 0 to node 9. Using the original Xmesh routing strategy, the packet will be forwarded to intermediate nodes with the best link quality from source node 0. Using the Xmesh based on the 21-node Cayley graph (Fig. 2), the packet will be sent to either node 3 or node 4, depending on the link quality between nodes 0 and 3, and nodes 0 and 4. Using the Xmesh based on random selection of routes, the packet will be randomly forwarded to any one of node 0's neighbors, i.e., nodes 3, 4, 18, or 11. We expect that, in general, messages routed according to Xmesh based on Cayley with shortest path routing to have the shortest path length, while that of random routing will impose longer path lengths.

These three routing strategies were simulated via Power TOSSIM for networks of sizes 21, 55, 110, 253, and 465 nodes. The simulation scenario is that each node continuously sends packets to the base station (node 0) at the transmission rate of 38.4 kbps with the packet size of 46 bytes. These numbers are used according to the specification of MICA2 motes by Crossbow Technology Inc.

By comparing these three strategies, we can see the effect of imposing a virtual topology on Xmesh. We expect the strategy of Xmesh based on Cayley with shortest path routing to have shorter path lengths than that of the original Xmesh. By imposing the Cayley topology with a random routing strategy, path lengths generated will be longer than the original than the original Xmesh.

All simulation were executed for the same amount of virtual/simulated seconds and with the same random seed and with the same random location deployment for networks of the same size regardless of their routing strategies. The radio model in Power TOSSIM sets the bit error rate between motes according to their location and various models of radio connectivity. We used the CC1000's "Empirical" radio model.

4.3 Power Consumption Analysis

To compare the power consumption required for the different strategies, Fig. 3 plots the averaged power consumption versus different network sizes. The averaged power consumption is the average of the total power consumption for all nodes in the network. Among the three strategies, the *random* routing strategy consumed the most power, while the topology-based (*Cayley)* on X-mesh consumed the least power. This is expected as the averaged path lengths for the *Cayley*

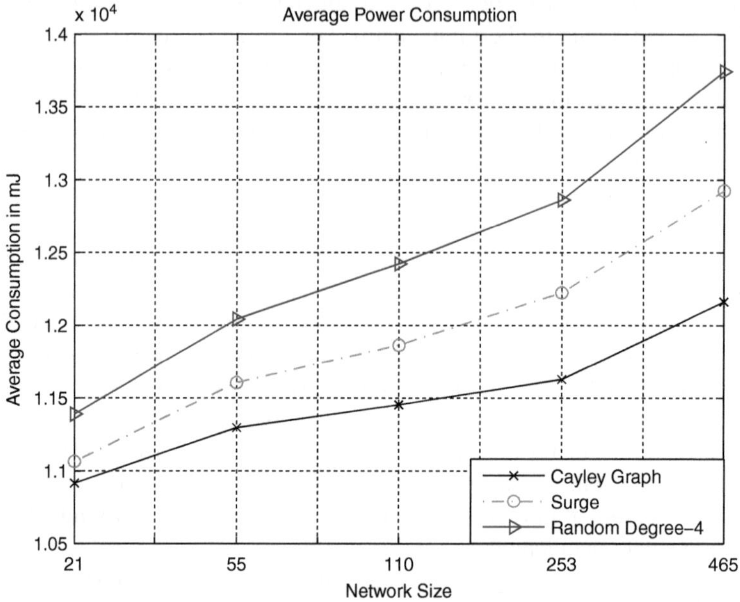

Fig. 3 Average power consumption for different routing strategies

	21-node	55-node	110-node	253-node	465-node
Cayley	2.65	3.01	3.95	5.84	6.91
Surge	2.81	3.55	4.53	6.27	7.46
Random	2.93	3.84	4.98	6.93	8.04

strategy is the shortest and that of the *random* (Cayley based with random routing) is the longest.

As expected, the averaged power consumed grows with the size of the network. Furthermore, the differences among the three strategies also grow with the size of network. For the small size network with 21-node, Cayley and Surge are about the same. For the largest size network, 465-node, Cayely's averaged power consumed is about 93% that of the original Xmesh (*Surge*). The difference is a consequence of the path lengths of the three routing strategies. Figure 4 plots the averaged hop counts among the three strategies for the different size networks. Indeed, the topology-based with random routing (*Random*) has the largest hop count, whereas the Cayley with shortest path routing (*Cayley*) has the shortest hop count. Furthermore, such difference grows with the size of the networks.

To further analyze the power consumption among the three different strategies, Fig. 5 plots the histogram of the energy consumed for the 55-node network. From the histogram, we observe that the Cayley strategy has a wider energy distribution than that of the original Xmesh (*Surge*) and the *random* case. For Cayley, the range is from 10,000 to 13,000 mJ, whereas the original Xmesh has a narrow peak around

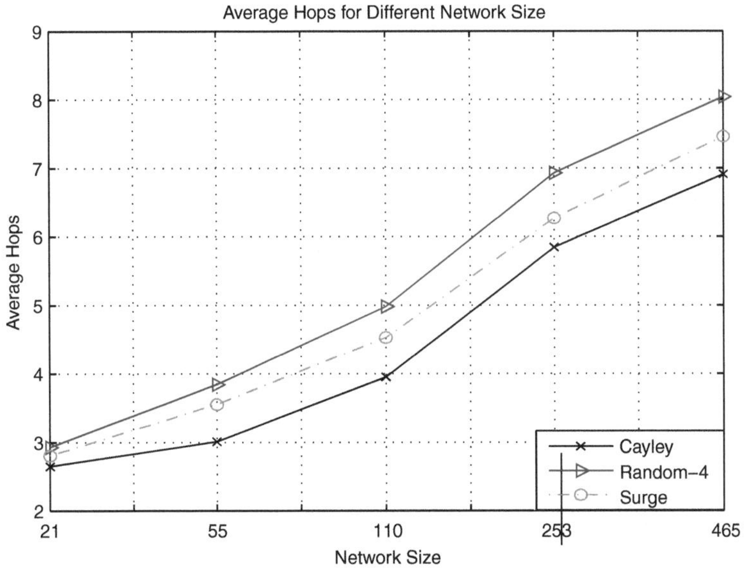

Fig. 4 Averaged hop count vs. network size

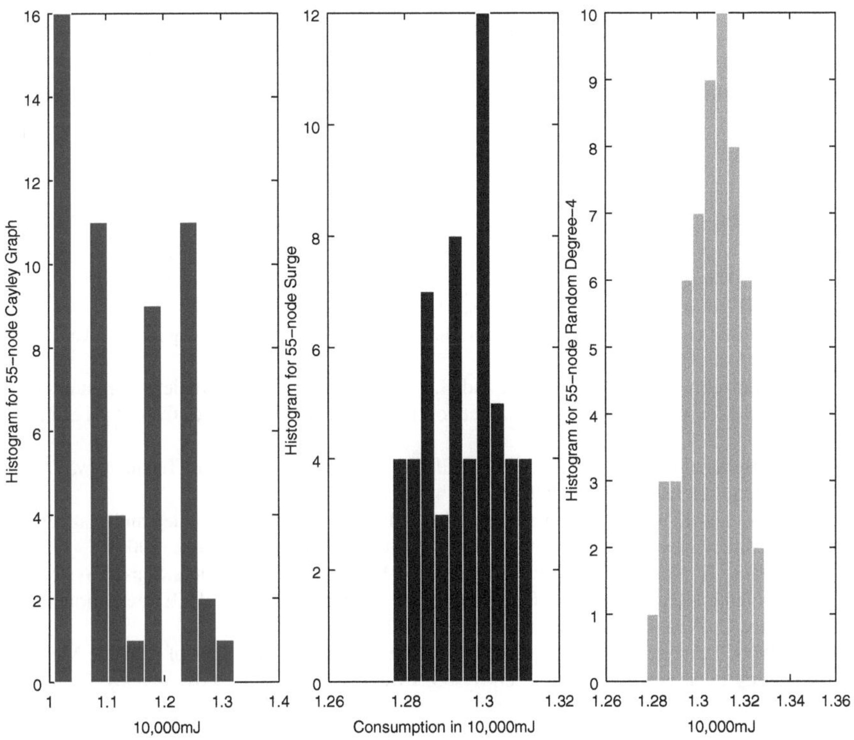

Fig. 5 Histogram of power consumption for 55-node network

13,000 mJ. As for the topology-based with random routing (Random), the range of the power consumption is between 12,700 and 13,300 mJ. Based on this result, we can conclude that the Cayley strategy conserved power but impose a wider energy distribution among nodes in the network. This implies that some nodes of the network may "die" first before others. To further investigate on this point, we computed the normalized standard deviation of energy consumption among nodes in a network for the three different routing strategies. Indeed, we found that the standard deviation for the topology-based routing with shortest path (*Cayley*) has a standard of deviation between 6 and 7% for different network sizes, whereas those of the Xmesh (*Surge*) and random (*Random*) routing is about 2%.

5 Conclusions

In this chapter, we have proposed a topology-based routing for Xmesh that combines the dense and vertex-transitive property of Cayley graphs. The dense property of Cayley graphs implies that path lengths are shorter and that the vertex-transitive property allows the same routing strategy and routing table be used at every node. Through the Power TOSSIM emulator, we implemented our proposed protocol for the Mica2 motes from Crossbow Technology Inc. [19]. Our simulation result for network size ranges from 21-node to about 500-node showed that the proposed topology-based routing consumes less power than the original Xmesh strategy. Furthermore, this power saving advantage of the proposed protocol increases with the network size.

References

1. J.N. Al-Karaki and A.E. Kamal, "Routing Techniques in Wireless Sensor Networks: A Survey", IEEE Wireless Communications, 11, 6, December 2004, 6–28.
2. I. Akyildiz et al., "A Survey on Sensor Networks", IEEE Commun. Mag., 40, 8, Aug. 2002, 102C14.
3. W. Heinzelman, A. Chandrakasan, and H. Balakrishnan, "Energy-Efficient Communication Protocol for Wireless Microsensor Networks", Proc. 33rd Hawaii Intl. Conf. Sys. Sci., Jan. 2000.
4. J.M. Hellerstein, W. Hong, and S.R. Madden, "The Sensor Spectrum: Technology, Trends, and Requirements", ACM SIGMOD 32, 4, December 2003, 22–27.
5. E. Noel and K.W. Tang, "Novel Sensor MAC Protocol Applied to Cayley and Manhattan", IEEE International Workshop on Wireless Ad-hoc and Sensor Networks, 2006
6. L. Wang, E. Noel, C. Fong, R. Kamoua, and K.W. Tang, "A Wireless Sensor System for Biopotential Recording in the Treatment of Sleep Apnea Disorder", IEEE International Conference On Networking, Sensing and Control, 2006.
7. A. Woo, T. Tong, and D. Culler, "Taming the Underlying Challenges of Reliable Multihop Routing in Sensor Networks", Sen-Sys03,November 5C7, 2003, Los Angeles, California, USA.
8. A. Woo and D. Culler, "Evaluation of efficient link reliability estimators for low-power wireless networks", Technical Report UCB//CSD-03-1270, U.C. Berkeley Computer Science Division, September 2003.

9. B.W. Arden and K.W. Tang, "Representations and Routing for Cayley Graphs", IEEE Transactions on Communications, 39, 11, November 1991, 1533–1537.
10. K.W. Tang and B.W. Arden, "Vertex-Transitivity and Routing for Cayley Graphs in GCR Representations", in the 1992 Symposium on Applied Computing, pp. 1180–1187, March 1–3, 1992, Kansas City, MO.
11. K.W. Tang, "Dense, Symmetric Interconnection Networks", Ph.D. Thesis, University of Rochester, 1991.
12. K.W. Tang and B.W. Arden, "Representations of Borel Cayley Graphs", SIAM Journal on Discrete Mathematics, 6, 4, November 1993, 655–676.
13. D.V. Chudnovsky, G.V. Chudnovsky, and M.M. Denneau, "Regular graphs with small diameter as models for interconnection networks", Tech. Rep. RC 13484–60281, IBM Res. Division, Feb. 1988.
14. C.L. Liu, "Shortest Path In Weighted Graphs", In Elements Of Discrete Mathematics, 2nd Edition, p. 147, McGraw-Hill, New York, 1998
15. D.B. Johnson and D.A. Maltz, "Dynamic Source Routing in Ad Hoc Wireless Networks" In T. Imielinski and H.F. Korth, Eds., Mobile Computing, vol. 353, Kluwer Academic Publishers, Dordrecht, 1996.
16. C. Ming-Syan and K.G. Shin, "Depth-First Search Approach for Fault-Tolerant Routing in Hypercube Multicomputers", IEEE Transactions On Parallel and Distributed Systems, 1, 2, April 1990, 152–159.
17. J. Silvela and J. Portillo, "Breadth-First Search and Its Application to Image Processing Problems", IEEE Transactions on Image Processing, 10, 8, August 2001, 1194–1199.
18. V. Shnayder, M. Hempstead, C.Bor-rong, and M. Welsh, "Simulating the Power Consumption of Large-Scale Sensor Network Applications", SenSys'04, November 3–5, 2004.
19. Crossbow Technology Inc's homepage, http://www.xbow.com, accessed in August 2007.
20. Crossbow, "Hardware Framework for Sensor Networks", In Presentation of Day one in Crossbow Seminar, Towson, December 2005.
21. Crossbow, "Multihop-Mesh-Network", In Presentation of Day one in Crossbow Seminar, Towson, December, 2005.
22. MoteIV Website, http://www.moteiv.com/, accessed in August 2007.
23. R. Fonseca, S. Ratnasamy et al., "Beacon-Vector Routing: Scalable Point-to-Point Routing in Wireless Sensor Networks", In Proceedings of NSDI 2005.
24. MICA2 Radio Stack for TinyOS, http://www.tinyos.net/tinyos-1.x/tos/platform/pc/CC1000Radio/CC1000RadioIntM.nc

Modeling Cell Placement and Coverage for Heterogeneous 3G Services

Roger M. Whitaker, Steve Hurley, and Stuart M. Allen

Abstract Locating cells for 3G services is a complex process due to the heterogeneity of service requirements. These requirements considerably affect the service coverage and resource utilisation of the network. This is a complex issue to model in WCDMA networks due to the dynamic nature of power control and the dependency between coverage and load. In this work we consider the issue of evaluating user service coverage and resource utilisation, or load that a trial candidate network design of cells can sustain. We investigate the sensitivity of two fundamental variables associated with snapshot evaluation – priorities for admission and the mix of requested services. This is conducted for the downlink scenario using a highly efficient offline evaluation technique that avoids recourse to lengthy simulation of online system power control. The results expose the significant influence that the variables have on the resultant evaluation of a candidate network design using a single traffic snapshot. This is important knowledge for network planning.

1 Introduction

The cellular concept has been used for a wide variety of fixed and mobile wireless services since the development of "radio" broadcasting for the mass population. This was closely followed by the introduction of terrestrial television and the use of private mobile radio (PMR) for emergency services. These provide 2-way communication using a single dedicated base station. A simple press-to-talk action is used

---

This is an extended manuscript based on "Sensitivity of service coverage evaluation for WCDMA systems", by R.M. Whitaker, S. Hurley and S.M. Allen which appeared in the Proceedings of the 2007 Wireless Telecommunications Symposium, Pomona, CA. ©2007 IEEE.

R.M. Whitaker (✉)
School of Computer Science, Cardiff University, Queens Buildings, 5 The Parade, Cardiff, CF24 3AA, UK
e-mail: R.M.Whitaker@cs.cf.ac.uk

S. Powell, J.P. Shim (eds.), *Wireless Technology*, Lecture Notes in Electrical Engineering 44, DOI 10.1007/978-0-387-71787-6_16,

which provides group listening and enables calls to be set up instantly. The first generation (1G) public cellular phone services were introduced in the 1980s with many then desirable features including the facility for a mobile to call from anywhere covered by a transmitter. In the 1990s, the second generation (2G) cellular systems, such as the European GSM, were responsible for bringing mobile communication to the wider general public. These digital systems originally offered voice and low data rate services with the option of GPRS functionality for packet-based data transfer.

The progression to third generation (3G) systems (such as UMTS and cdma2000) has been somewhat less rapid than development seen in the 1990s. But 3G services are expected to facilitate high data rate mobile services (up to 2 Mbps) such as video on demand and web browsing. However, parallel developments such as the introduction of wireless local area networks (WLAN) and selective use of wide area data networks offer potential alternative solutions for indoor applications.

The progression to mobile high data rate services has sought to blur the previously clear distinction between mobile communication, computing and telephony. Many different types of multimedia services are characterised only by bit rate, quality of service and user mobility requirements. Transmission then takes place using a unifying wireless protocol such as wide-band CDMA in the case of 3G. The cellular approach remains central to the provision of these services and consequently planning methodologies are most important. The issue of planning, in terms of transmission infrastructure and configuration, is needed to address quality of service, taking cost–effectiveness into account. There are a large number of complex engineering factors affecting such network design. These include choice of antennae, power control, tilt and azimuth settings as well as alternative choices for location of equipment. Such factors result in a high complexity problem which is unlikely to be solved to (near-) optimality without appropriate modelling

This has driven the development of computational techniques and software to optimise network deployment [19]. The benefits of such techniques are two-fold: first, they ensure that the network is designed to cope with the coverage and capacity requirements of potential users; second, they ensure that the operator′s fixed costs, primarily infrastructure investment, are kept as low as possible. These are conflicting objectives that require careful bilateral analysis [21]. Fundamental required functionality in network design is the ability to evaluate a candidate network design with reference to cost, user coverage and capacity (or *load*). A network design constitutes a configured selection of located transmitter sites. Evaluation requires situated modelling of the communication system at the physical layer.

To achieve this a range of detailed models were originally proposed, particularly for frequency–time division access systems such as second generation GSM as in [6, 9, 10, 11, 15, 16, 18]. In most cases, these models independently address service coverage and capacity. This is highly convenient for modelling purposes and it has led to abstract set covering-based approaches. However, this is unsuitable for third generation (3G) cdma2000 and UMTS systems based on WCDMA. This is because in these systems, there is a close coupling between coverage and capacity which requires link level detail. This can be counter productive to the process of optimised cell location as it means that there is potentially a much greater computational burden on evaluating candidate solutions. Although a number of models have been proposed for optimisation of UMTS network design [1–3, 7, 11, 14, 17], these have been based on widely varying assumptions and degrees of abstraction. Our purpose in this chapter is to closely analyse and address the important issue of evaluating, with detail, the coverage in CDMA–based systems as used for 3G services.

2 Evaluating Coverage in CDMA-Based Networks

A fundamental requirement in network design is that of being able to evaluate a candidate network design and interpret the associated sensitivity. This is easily achieved when modelling FDMA-based systems. However, coverage and capacity evaluation in CDMA-based systems such as UMTS are intimately linked. This is best explained by considering the downlink scenario, where proportion of total transmission power utilised at a cell represents its *load*. However, the amount of power allocated to support an individual link depends on the amount of interference present at the mobile receiver. This in turn depends on the power of transmissions supporting links to other users since the power used to support a link to an intended particular receiver is seen as interference by all other receivers. In operational systems such as UMTS, this is managed by fast power control which makes frequent per-link power adjustments (circa 1,500 Hz) in response to achieving target signal-to-noise and interference ratio (SNIR).

Simulating per-link power control is not an efficient approach to determining coverage and load since every time a power allocation is adjusted, SNIR needs to be re-evaluated by all other users and power adjustments made accordingly. This needs to be repeated until stability is attained. Furthermore, such power control simulation is usually performed on different network *snapshots*, which are static spatial distributions of simultaneously active users with particular requests for heterogeneous services. Rapid evaluation of service coverage and load is very desirable so that network performance relative to many different snapshot scenarios is possible.

There are some contributions that assist offline evaluation. These are mostly analytical and based on closed-form expressions of system variables. When a network is known a priori to have sufficient capacity to serve all users, solution methods are known (e.g. [13]) to exactly determine the minimum transmit and received powers. However, sufficient capacity for all users is not a realistic underlying assumption when planning a network. Further analytical contributions [4] have developed a technique to identify cells that are unable to support all downlink-offered traffic, based on a mixed integer mathematical programme to find a suitable scaling vector. In contrast we adopt an algorithm which is particularly detailed and flexible [20]. This approach is particularly of low complexity (linear) and permits radio resource management policies, extensible to soft-handover, to be directly incorporated in resource constrained planning scenarios. Its low complexity means that typically less than 5% of computation time, as compared to simulation of power control, is required. This approach has also been shown [20] to closely approximate the coverage and loading characteristics as seen from simulated power control.

The computational approach can be used to explore the sensitivity of downlink snapshot evaluation and the extent to which changes in admission policy and requested services affect system loading and user coverage. This assessment is crucial because it affects the perceived quality of a candidate network configuration, as interpreted by a radio-planner or an automated system for network planning. Our experiments fix the geographical locations at which users may be located and observe the effects of varying the requested services and priorities for user admission within the model. This permits insights into the extent to which higher data rate services and admission assumptions affect snapshot evaluation.

2.1 Downlink Model

The parameters used are described in Table 1. We use the terms *cell* and *antenna* interchangeably since each cell is served by a single antenna and assume that antennae may be co-sited at a base station location. Base stations with particular antenna configurations (tilt, number of co-sited antennae, direction and height) are assumed across a given planning region. Discrete locations from this region called *test points*

Table 1. Downlink parameters

Parameter	Description
W	CDMA chip rate
R_i	Data rate for service *i*
S_{ptp}	The set of covered pilot test points
S_{stp}	The set of service test points
O_{stp}	An ordering of the *stp*
I_{own}	Total power received from serving cell (all links and pilot)
I_{oth}	Total power received from all other cells than the serving cell
P_n	Noise power seen at a test point (thermal and equipment)
α	Orthogonality factor
P_{xy}	Power allocated by cell y for *stp* x
p_{xy}	Power allocated by cell y for *stp* x as received at *stp* x
PL_{xy}	inclusive total path loss (dB) between *stp* x and cell y
p_{xy}^{pilot}	Pilot power from cell y as received at *stp* x
$(E_c\, I_o)_{pilot}$	Target threshold for pilot E_c/I_o ratio
$(E_b\, N_o)^*_{DL}$	Target threshold for E_b/N_o ratio for the dedicated DL channel (dependent on required service)
$\eta_{DL,y}$	Downlink load at cell y
$\eta_{DL,y}^{\max}$	Maximum allowed downlink load at cell y
$Ptxtotal_y$	Actual total allocated traffic transmission power in cell y
$Ptxtotal^*_y$	Assumed total allocated traffic transmission power in cell y
$Ptx\max_y$	Maximum transmit power capability of cell y
$\eta_{pilot,y}$	Proportion of $Ptx\max_y$ allocated for pilot signal at cell y

are used to spatially sample service coverage. Path loss data are required to assess the received signal strength. This may be estimated using empirical models or derived from data sampled in the field. Two types of test point are defined in our model: *service test points* (*stp*) and pilot test points (*ptp*). The *ptp* are used to assess *pilot* signal quality, the adequate downlink reception of which is a necessary prerequisite for system control. At an *stp*, quality of downlink dedicated traffic channels is assessed for a particular service, which is defined prior to snapshot evaluation.

2.2 Test Point Coverage and Cell Load

The pilot signal is transmitted at a fixed proportion $\eta_{pilot,y}$ of maximum cell power. A *ptp* x is *served* by antenna y when the received energy per chip relative to total the total power spectral density E_c/I_o at least meets the required target threshold $(E_c/I_o)_{pilot}$. Letting

$$I_y = I_{own} + I_{oth} \tag{1}$$

then x is *served* if and only if

$$\frac{p_{xy}^{pilot}}{N + I_y} \geq (E_c/I_o)_{pilot} \tag{2}$$

An *stp* is *covered* in the downlink direction if energy per bit relative to spectral noise density E_b/N_o at least meets the required target threshold. Specifically, for an *stp* x and serving antenna y, x is *DL covered* if and only if

$$\frac{W}{R_i} \cdot \frac{p_{xy}}{I_{own}(1-\alpha) + I_{oth} + P_n} \geq (E_b/N_o)^*_{DL} \tag{3}$$

There are various ways in which system resource utilization, or cell loading, can be assessed but in this instance it is convenient to use wideband power-based measurement because it directly identifies with the resources being allocated. The downlink load at cell y is estimated by

$$\eta_{DL,y} = \frac{PtxTotal_y}{Ptx\max_y} \tag{4}$$

Note that *ptp* coverage is dependent on cell load, which governs the amount of interference received. Therefore, we define a *ptp* as *covered* if and only if it is *served* when all cells y are operating at maximum permitted downlink load $\eta_{DL,y}^{\max}$. Consequently covered *ptp* can see the pilot signal independent of traffic and are collectively denoted S_{ptp}. This set is identifiable a priori and to ensure that an *stp* can see the pilot signal, it is required that $S_{stp} \subseteq S_{ptp}$.

To constrain overloading, it is necessary to identify the relative importance of an *stp* for potential admission to the system. We account for this using a list O_{stp} of the set S_{stp} specifying the prioritisation for admission. A *traffic snapshot* is then defined as the triple $\{(S_{ptp}, S_{stp}, O_{stp})$. Such snapshots are used to assess a networks performance relative to different spatial traffic requirements and many such snapshots may be employed in practice. As indicated in [8], traffic snapshots may be sampled from Monte Carlo traffic simulation and are the most common approach to off-line physical layer network evaluation. Our attention in this chapter concerns evaluation relative to a single snapshot.

2.3 Algorithmic Approach

We apply the algorithm developed in [20] using a single server model for downlink snapshot assessment. This is the limiting case for higher data rate services. We explain the algorithm using five steps which constitute a single iteration and for demonstration purposes, we assume that each *stp* is only potentially admissible by its best (i.e. least path loss) server. At each iteration *stp* are admitted to the network, by the algorithm, if the *stp's* best server has sufficient resources to meet the *stp's* required target E_b/N_o ratio (equivalently SNIR) given the level of interference from *assumed* cell transmission powers. These *assumed* transmission powers are updated at each iteration until they represent those actually allocated to admitted users. The six steps for a single iteration of the algorithm are as follows:

1. For each cell y set the *assumed* total allocated traffic transmission power $Ptxtotal_y^*$.
   - At the first iteration, this is maximum cell load, typically circa 0.6, and this results in $Ptxtotal_y^* = 0.6 \cdot Ptx\max_y$.
   - At subsequent iterations, $Ptxtotal_y^* = Ptxtotal_y$.
2. For each cell y, set the *actual* total allocated traffic transmission power $Ptxtotal_y$ as zero.
3. For each *stp* x with best server denoted y, on rearranging Eq. (3) determine p_{xy} where

$$p_{xy} = (E_b \,/\, N_o)_{DL}^* \cdot \frac{R}{W} \cdot (1-\alpha)I_{own} + I_{oth} + P_n \tag{5}$$

4. Also determine P_{xy} where

$$P_{xy} = p_{xy} \cdot 10^{PL_{xy}/10} \tag{6}$$

5. Taking the *stp* as ordered in the snapshot list O_{stp} , each *stp* is sequentially considered for admission to its best serving cell. An *stp* x in the sequence O_{stp} is admitted to its best serving cell y if $P_{xy} + Ptxtotal_y \leq Ptxtotal_y^*$ in which case the addition of P_{xy} to the current *actual* transmission power of cell y does not

cause the *assumed* total transmission power to be exceeded. This ensures that the interference levels I_{own} and I_{oth} in step 3 do not underestimate (but could overestimate) actual transmission powers across the network. If x is admitted to cell y, $Ptxtotal_y$ is updated accordingly, i.e. $Ptxtotal_y \rightarrow Ptxtotal_y + P_{xy}$.
6. For all cells y, set $\varepsilon_y = Ptxtotal_y^* - Ptxtotal_y$.

On completion of one iteration, the conservative error in the algorithm is the difference between the assumed and the actual total traffic transmission power levels, indicated by ε_y for each cell y. This is reduced at the next iteration by updating the assumed traffic transmission power levels in step 1, resulting in $\varepsilon_y \rightarrow 0$ as the number of iterations increase. Moreover, since for all cells y, $Ptxtotal_y$ is less than or equal to its value at the previous iteration, the power p_{xy} required for coverage of *stp* x is also less than or equal to its value at the previous iteration. The net effect is equal or increased *stp* coverage at subsequent iterations. The algorithm may terminate after a fixed number of iterations or after criteria on max (ε_y) $ are satisfied.

2.4 Complexity

Each step of the algorithm involves sequential operations on the number of *stp* or cells. This results in linear complexity, a substantial reduction on that seen for simulation of online power control, where on admission of an additional *stp*, all previously admitted *stp* re-evaluate their E_b/N_o attainment and adjust power accordingly. This results in $O(n^2)$ complexity for simulation of online power control assuming n *stp*.

3 Demonstration of Evaluation

Our demonstration shows some of the underlying sensitivities in WCDMA snapshot evaluation. Previously, the overhead from detailed simulation of online power control has impeded their exposition in a large-scale wide-area setting. The proposed approach resolves this while being a sound comparative technique.

3.1 Test Problems

We adopt a 20 km by 20 km transmission region as shown in Fig. 1. This consists of 20 transmission sites (white pixels) each hosting three cells configured with a wide beam antenna sited at 0, 120 and 240° intervals in the horizontal plane. Each has a 5° down-tilt in the vertical plane.

Static path loss predictions have been obtained from a Hata-based model [5] with additional random fluctuations to further simulate random effects such as fading. Directional effects of antennae relative to the locations of the test points are

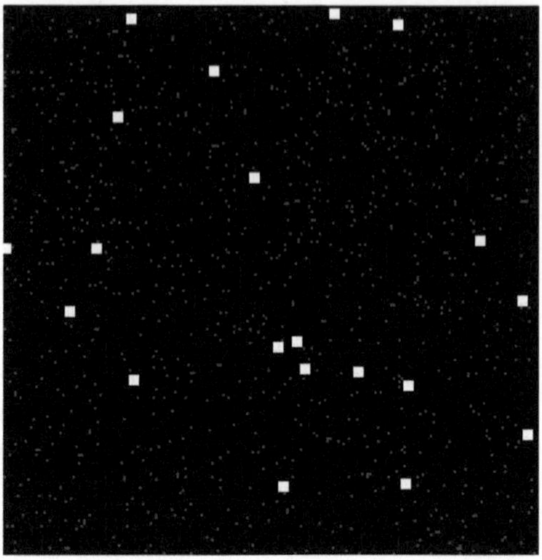

Fig. 1 View of transmission locations (*white*) and *ptp* locations from which *stp* snapshots are selected (*ptp* indicated by small pixels)

incorporated in PL_{xy} values. Path loss estimates are included up to a maximum of 255 dB.

A set of 1,000 fixed location *stp* are used for experimentation. These are indicated in Fig. 1 by the small highlighted pixels. These are randomly selected and form the basis for defining *stp* locations throughout. For demonstration purposes, we consider the impact of varying (i) the order O_{stp} and (ii) the services *stp* request. To achieve (i), a common set of 100 random *stp* orderings have been defined, along with best-server-first and best-server-last orderings. To achieve (ii) the *stp* request four different services $R_1, \ldots, R_4$ as characterised in Table 3. Four data rates are employed for services $R_1, \ldots, R_4$ as indicated in Table 2.

Table 2 Variable settings for experimentation

Parameter	Setting
W	3,840,000 chips/sec
R_i	12.2,64,144,384 kbps for service $R_1, \ldots, R_4$
$(E_b/N_o)^*_{DL}$	6.5,6.0,5.0,4.5 dB
$P_n, \alpha, \eta^{\max}_{DL,y}$	–108 dBm, 0.5, 0.6
$(E_c/I_o)_{pilot}$	–18 dBm
$Ptx\max_y$	43 dBm

3.2 Algorithm Behaviour

Figures 2 and 3 show the effect of algorithm iteration on the average coverage and load. While modest changes (increases) are seen in coverage as iterations increase, more significant are seen in actual/assumed load. In particular Fig. 3 shows the

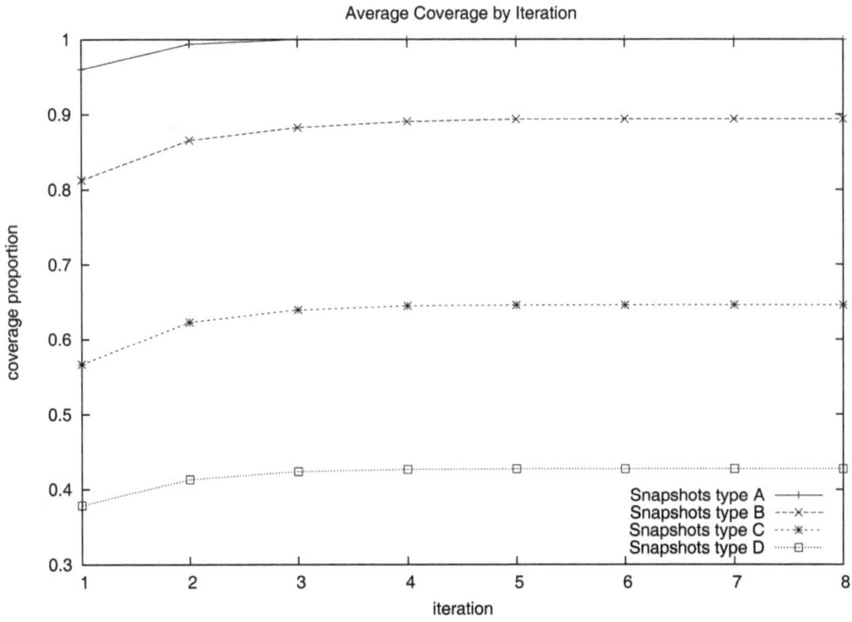

Fig. 2 Average coverage by iteration for different samples of snapshots

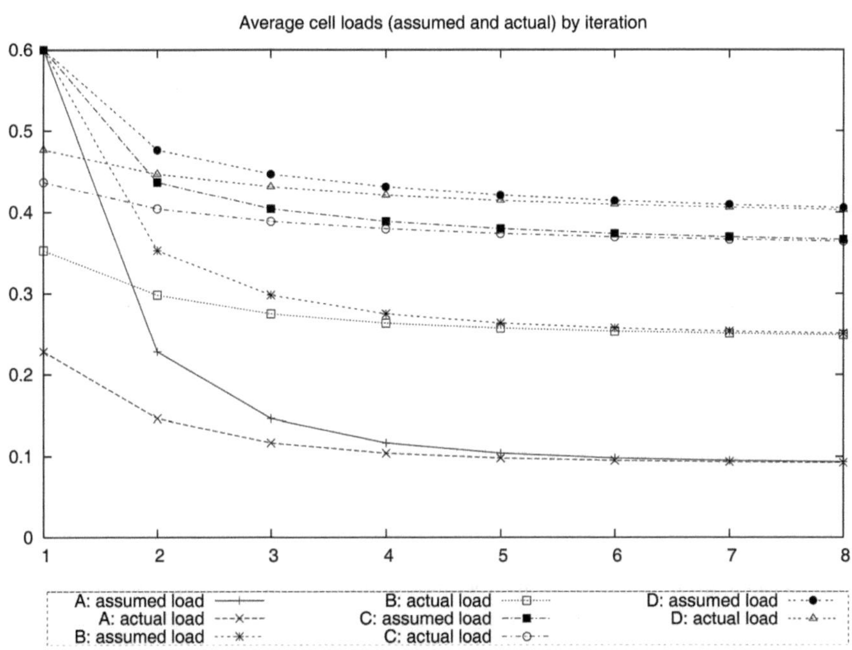

Fig. 3 Average cell load (assumed and actual) by iteration, for snapshot samples *A, B, C, D*

Table 3 Composition of snapshot samples

Snapshot index	Data rate 12.2 kbps	Data rate 64 kbps	Data rate 144 kbps	Data rate 384 kbps
A	100%	0%	0%	0%
B	75%	25%	0%	0%
C	50%	25%	25%	0%
D	25%	25%	25%	25%

average convergence of η_y, which is rapid for the first few iterations. From this it is evident that overall, fewer iterations are required to approximate coverage than for load.

3.3 Services Requested and User Coverage

The services that users request significantly affect the networks ability to simultaneously serve users. Figure 2 indicates the average number of simultaneous users, over the same 100 admission orderings for each of the service mixes in Table 3. Higher data rate users are challenging to satisfy and this is achieved at the expense of multiple low data rate users. This is shown in Table 4 where the average percentage of satisfied users, classified by service, is shown for the 100 snapshot samples of types $A,\ldots,D$. It is clear that high data rate services are significantly more difficult to satisfy. For example, in snapshots of type D, on an average 18.4% of users satisfied are low data rate, while only 4.8% of users satisfied request the highest data rate.

Table 4 Percentage of *stp* covered by service

Snapshot sample	Data rate R_1	Data rate R_2	Data rate R_3	Data rate R_4
A	98.0%	0%	0%	0%
B	68.2%	21.1%	0%	0%
C	37.8%	14.8%	11.9%	0%
D	18.4%	11.2%	8.3%	4.8%

3.4 Effect of Priorities for Admission and User Coverage

Figure 8 shows the average coverage from sampling 100 snapshots compared with least-path-loss first and greatest-path-loss first snapshot orderings. As higher data rate services are introduced (snapshots type C and D), there is increased variability on the number of *stp* covered due to *stp* ordering. Note that due to increased path loss, *stp* being far from their best server require more resources to meet target SNIR and the consequences of this become compounded when higher data rate services, with increased resource requirements over low data rate services, are requested.

However, it is noteworthy that the greatest-path-loss first snapshot ordering is a useful single conservative ordering to adopt.

3.5 Services Requested and Load

Figures 4, 5, 6 and 7 show average load by cell for each of the 100 snapshot samples of types *A, B, C, D*, respectively. The uneven loading distributions seen across cells are a function of the randomised network design which is not optimised in terms of site location or configuration. It is noticeable that those cells which remain under-capacitated generally do so across all other snapshot types. Clearly this evaluation is useful for informing changes to the network, either manually or via artificial intelligence-based network design approaches. Note that the final average cell loads for each snapshot sample type are shown in Fig. 9 and are, respectively, 0.09, 0.25, 0.36 and 0.41. This occurs while average total coverage decreases over snapshot types *A* to *D* as shown in Fig. 8.

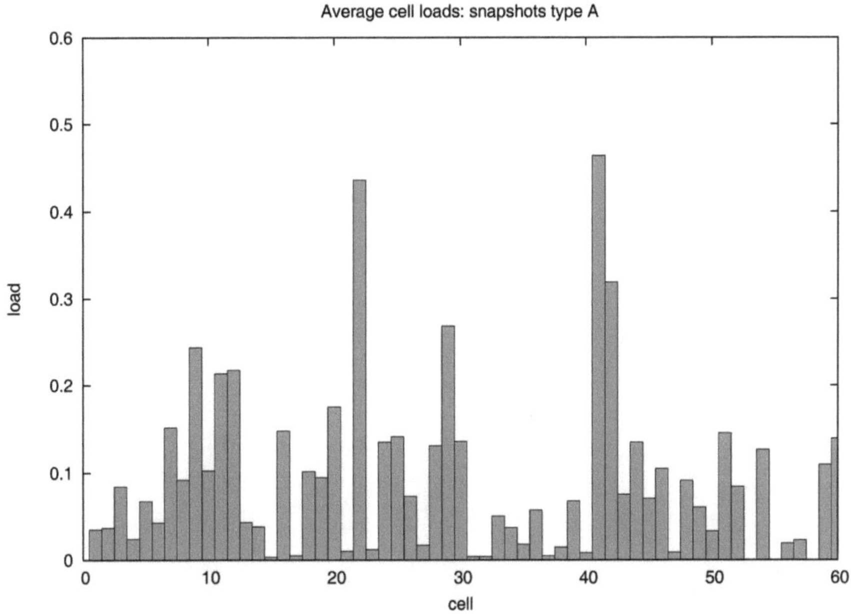

Fig. 4 Average cell loads for snapshots type *A*

3.6 Summary

Looking at detailed evaluation, it has been shown that WCDMA systems are highly sensitive to parameters that are set external to the analysis, such as *stp* ordering.

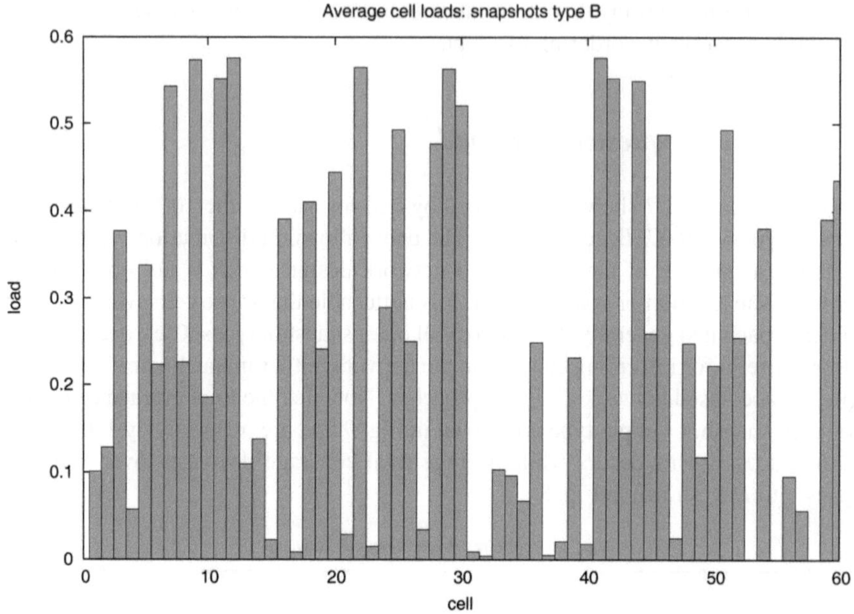

Fig. 5 Average cell loads for snapshots type *B*

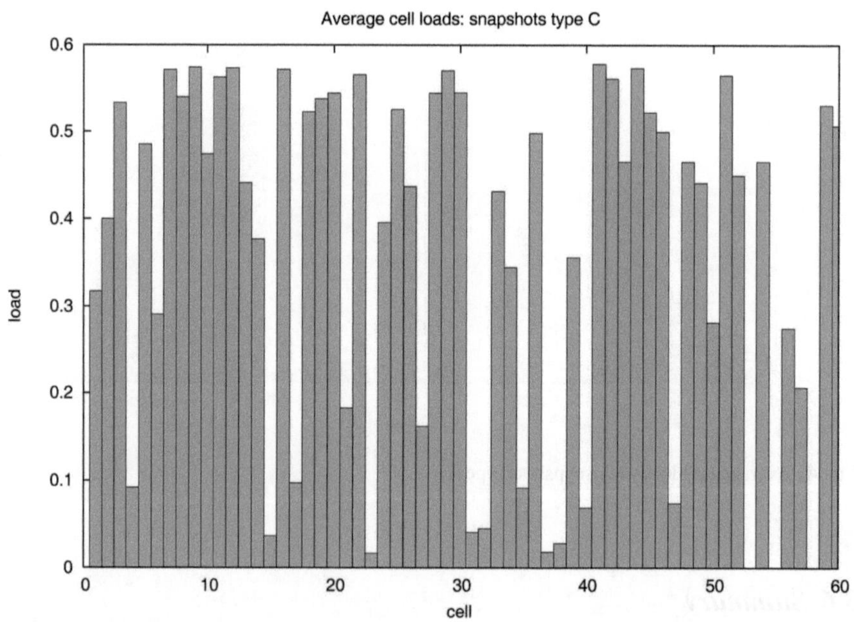

Fig. 6 Average cell loads for snapshots type *C*

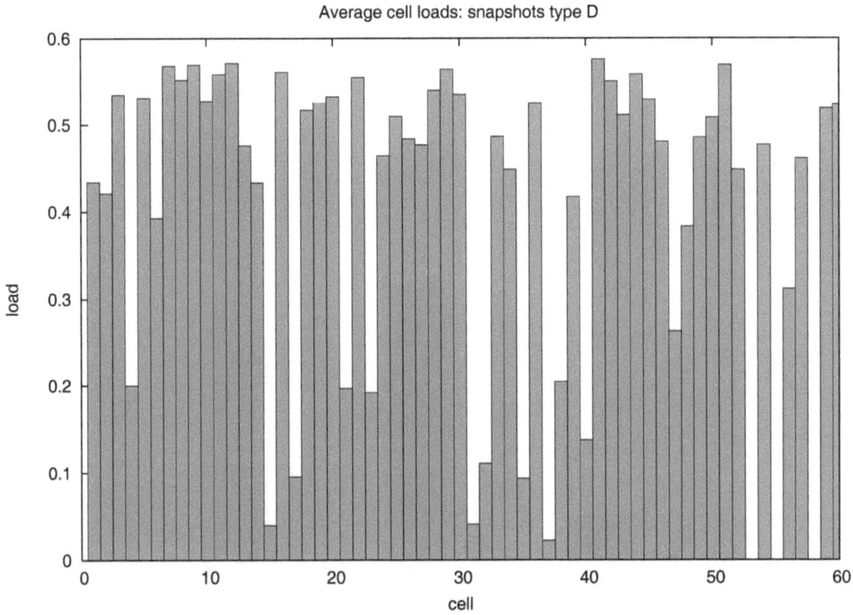

Fig. 7 Average cell loads for snapshots type D

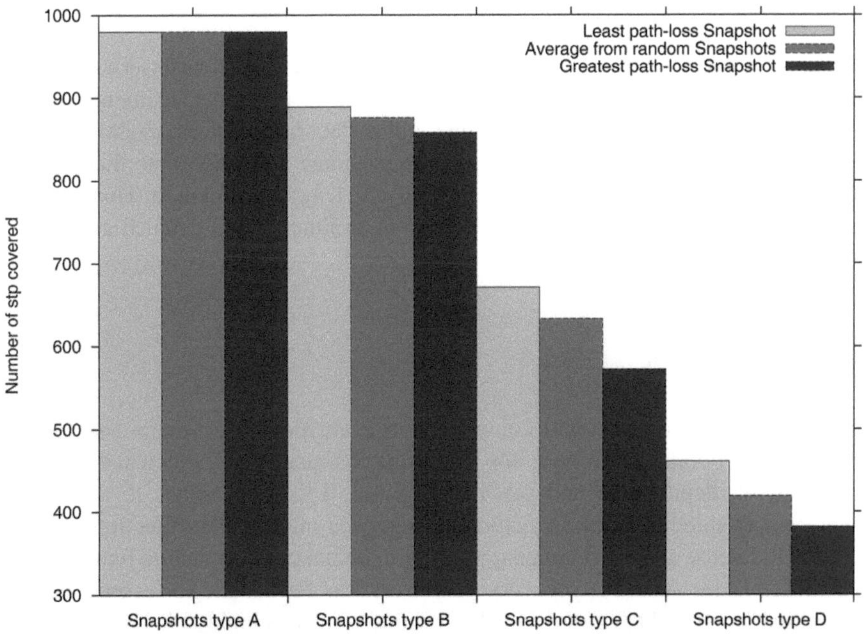

Fig. 8 Average coverage from sampling as compared with least-path-loss first and greatest-path-loss first snapshot orderings

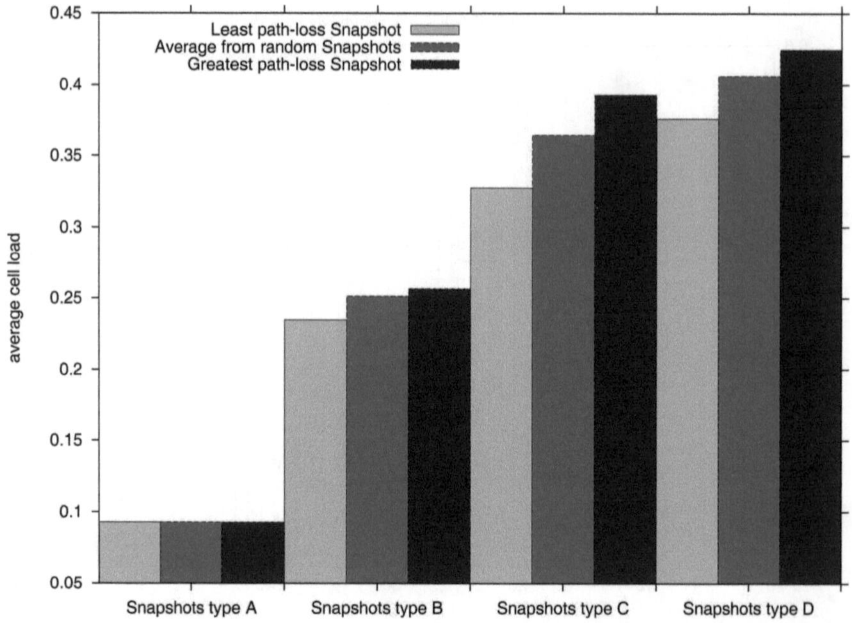

Fig. 9 Average cell load from sampling as compared with least-path-loss first and greatest-path-loss first snapshot orderings

However, greatest-path-loss first admission is a useful basis for conservative planning in the absence of knowledge on actual radio resource management policies. The effects of satisfying users requesting higher data rate services, as shown in this chapter, are also significant and affect both coverage and load. Note that this has been examined in this chapter without recourse to lengthy simulation. This is highly desirable functionality enabling rapid network evaluation and detailed assessment of sensitivity.

4 Future Trends

Solving the problem of rapid computational evaluation of coverage and load in WCDMA networks is the basis for increasing the quality and efficiency of future 3G network deployment and service provision. This also relates to the use of increasingly intelligent infrastructure and resource management functionality that is able to locally adapt to changing patterns in demand and changing requirements for different services. As yet, very little work has been done to incorporate this increased functionality at the network-planning stage. It is highly likely that increasingly sophisticated and dynamic cellular transmission equipment, such as steer-able antenna, will require detailed physical layer models. Incorporation of the potential behaviour of transmission infrastructure with reference to external conditions on mobility and user service requirements will also be necessary. This represents a future trend toward increased integration of artificial intelligence across multiple layers. This increasingly blurs the boundaries between traditionally distinct areas of system operation and emphasises the need for an increasingly integrated treatment of radio resource management and system deployment.

5 Conclusions

Cell placement and configuration is a crucial aspect of mobile service provision and represents the crucial point at which the service providers and wireless consumers interact. This issue is particularly important for 3G service which is resource intensive and consequently highly susceptible to poor network planning and cell placement. Developing computational techniques to support rapid and detailed cell-planning analysis is central to efficient deployment of 3G services. In this chapter we have examined network evaluation and demonstrated a technique to facilitate rapid coverage and capacity evaluation. In doing so, important sensitivities in network planning have been exposed.

Acknowledgements This research is supported by EPSRC grant EP/E020720, titled "Bounds for site selection and configuration in cellular networks."

References

1. Amaldi E, Belotti P, Capone A, Malucelli F (2006) Optimizing base station location and configuration in UMTS networks. Annals of Operations Research 146:135–151
2. Amaldi E, Capone A, Malucelli F (2001) Planning UMTS base station location: optimization models with power control and algorithms. IEEE Transactions on Wireless Communications 2(5):939–952
3. Eisenblatter A, Fugenschuh A, Geerdes HF, Junglas D, Koch T, Martin A. (2004) Integer programming methods for UMTS radio network planning. Proceedings of the WiOpt′04 Conference, Cambridge UK, 527–537
4. Eisenblatter A, Geerdes, HF (2005) Analytical approximate load control in W-CDMA radio networks. Proceedings of the IEEE 62nd Vehicular Technology Conference (fall), 1534–1538
5. Hata MH (1980) Empirical formula for propagation loss in land mobile radio services. IEEE Transactions on Vehicular Technology 29(3):317–323
6. Hurley S (2002) Planning effective cellular mobile radio networks. IEEE Transactions on Vehicular Technology 51(2):243–253
7. Jamaa SB, Altman Z, Picard JM, Fourestie B, Mourlon J (2004) manual and automatic design for UMTS networks. Mobile Networks and Applications 9(6):619–626
8. Laiho J, Wacker A, Novosad T (2002) Radio Network Planning and Optimization for UMTS, Wiley, Chichester
9. Lee CY, Kang HG (2000) Cell planning with capacity expansion in mobile communications: a tabu search approach. IEEE Transactions on Vehicular Technology 49(5):1678–1691
10. Mathar RM, Niessen T (2000) Optimum positioning of base stations for cellular radio networks. Wireless Networks 6:421–428
11. Mathar RM, Schmeink M (2001) Optimal base station positioning and channel assignment for 3G mobile networks by integer programming. Annals of Operations Research 107:225–236

12. Mathar RM, Schmeink M (2002) Integrated optimal cell site selection and frequency allocation for cellular radio networks. Telecommunication Systems 21:339–347
13. Mendo L, Hernando, JM (2001) On dimension reduction for the power control problem. IEEE Transactions of Communications 49(2):243–248
14. Molina A, Athanasiadou GE, Nix AR (2000) Optimised base-station location algorithm for next generation microcellular networks. Electronics Letters 26(7):668–669
15. Raisanen L, Whitaker RM (2005) Comparison and evaluation of multiple objective genetic algorithms for the antenna placement problem. Mobile Networks and Applications 10:79–88
16. Reininger P, Caminada A (2001) Multicriteria design model for cellular network. Annals of Operations Research 107:251–265
17. Tcha D.-W, Myung Y.-S (2000) Base station location in a cellular CDMA system. Telecommunication Systems 14:163–173
18. Vasquez M, Hao J.-K (2001) A Heuristic approach for antenna positioning in cellular networks. Journal of Heuristics 7(5):443–472
19. Whitaker RM, Hurley S (2003). Evolution of planning for wireless communication systems. Proceedings of 36th Annual Hawaii International Conference on Systems Sciences 387–391
20. Whitaker RM, Hurley S, Allen SM (2007) Power control Heuristics for efficient load evaluation in WCDMA Network Modelling. Submitted for publication.
21. Whitaker RM, Raisanen L, Hurley S (2005) The infrastructure efficiency. Computer Networks 48:941–959

Index

Note: Page reference with *f* and *t* notation refer to a *figure* and *table* on that page respectively.

MIX
Papier aus verantwortungsvollen Quellen
Paper from responsible sources
FSC® C105338
FSC
www.fsc.org

If you have any concerns about our products,
you can contact us on
ProductSafety@springernature.com

In case Publisher is established outside the EU,
the EU authorized representative is:
Springer Nature Customer Service Center GmbH
Europaplatz 3, 69115 Heidelberg, Germany

Printed by Libri Plureos GmbH
in Hamburg, Germany